D0983725

Relativity, Mechanics and Statistical Physics

D.S. Mann
and
P.K. Mukherjee
Department of Physics
Deshbandhu College
University of Delhi

A HALSTED PRESS BOOK

John Wiley & Sons

New York   Chichester   Brisbane   Toronto

Randall Library UNC-W

Copyright © 1981, Wiley Eastern Limited
New Delhi

Published in the Western Hemisphere by
Halsted Press, a Division of
John Wiley & Sons, Inc., New York

Library of Congress Cataloging in Publication Data

Mann, D.S.
    Relativity, mechanics, and statistical physics.

    " A Halsted Press book."
    Includes index.
    1. Relativity (Physics)   2. Mechanics.   3. Statistical
physics.   I. Mukherjee, P.K.   II. Title.

QC173.55.M36   1981   530.1' 1   81—4688
ISBN   0—470—27181—7   AACR2

Printed in India at Prabhat Press, Meerut.

Randall Library UNC-W

QC173
.55
.m36
1981

To

*Dear students
who inspired us
to write this book*

215482

Preface

At a time when text books in physics are emerging in endless procession, the authors of a new text are obliged to justify their efforts either in terms of contributions to a better understanding of the subject or to the advancement of the frontiers of scientific knowledge.

This book is designed to meet the requirements of B. Sc. students of Indian Universities. The syllabi, as revised by the universities in the wake of $10 + 2 + 3$ system, are off the beaten track and require a new approach towards the presentation of the subject. The existing books automatically have been rendered inadequate, and this has inspired the authors to write a new book. The book, the authors believe, will serve as a compact text book with an easier approach to the subject. It will be found especially useful for the B. Sc. (Pass) first year students of Delhi University. Some of the salient features of the book are:

1. It endeavours to present a reasonable, logical and systematic exposition of the topics on Relativity, Mechanics and Statistical Physics. It is hoped that this will enable the students to acquire a fair knowledge of the basic concepts of these topics and will serve as a springboard to attain a higher degree of proficiency in these topics. Throughout the book the emphasis has been on giving a clear, compact, logically correct and systematic account of the text.

2. The various topics have been treated with sufficient depth. The subject matter has been developed methodically. Long discussions and irrelevant details have been avoided and preference has been given to an analytical and explanatory approach. Evolution of the various topics is smooth and continuous.

3. In tune with the recent trend, SI system of units has been

mainly adopted. The book is suitably illustrated by simple and carefully designed diagrams. Captions have been provided with the illustrations which make them self-explanatory.

4. The necessary mathematics is explained step by step, and no gaps have been left for expert readers to fill and less experienced to neglect. However, it has been assumed that the students using this book will be equipped with a working knowledge of vectors, differential and integral calculus.

5. Selected numerical problems (with their answers) appended at the end of every chapter immensely increase the utility of the book. Hints have been provided with problems which are considered to be little beyond the comprehension of the average students.

6. To authors' knowledge, there is no single text book (Indian or foreign) which under one cover treats all the three topics included in the present book. To get the subject matter contained in the various topics of this book, one has to consult a number of different books by Indian as well as foreign authors. Unfortunately the books by Indian authors give a very sketchy treatment of the topics especially on Relativity and Statistical Physics. It is, therefore, felt that the book will fulfil the long-felt need of perplexed and harassed students to have a compact text book that may cater to their needs in keeping with the newly revised syllabi.

The book running through nine chapters is divided into three parts. Part One deals with the theory of Relativity and is spread over three chapters. Chapter 1 gives an account of the classical Newtonian mechanics starting with Galilean transformations. Einsteinian relativity and some consequences of the theory are treated in Chapter 2 while Chapter 3 is devoted to a discussion concerning relativistic dynamics.

Mechanics forms the subject matter of Part Two which comprises four chapters. Chapter 4 discusses the rotatory motion while Gravitation is dealt with in Chapter 5. Harmonic oscillator and its applications appear in Chapter 6 while elasticity and viscosity are discussed in Chapter 7.

Statistical physics is dealt with in Part Three of the book. This part is divided into two chapters. In Chapter 8 classical Maxwell-Boltzmann statistics and its applications are included, while the transport phenomena in gases and Brownian motion are given consideration in Chapter 9.

The best learning and teaching, the authors believe, can be achieved only when there is a direct relationship between the teacher and the taught. In that atmosphere new ideas can be provoked, discussed and understood. But this ideal situation has become impossible because of multiplying number of students in class rooms. Hence there arises the need to find a near substitute to that ideal. Perhaps our book may make some contribution to that end, and may inspire and enable the students to have a better and firm grasp of the topics discussed in the text.

We have great pleasure in extending our sincere and heartfelt thanks to the staff of Wiley Eastern Limited.

One of us (PKM) considers his sincere duty to place on record the silent involvement of his venerable parents whose blessings have always been the guiding force and a source of constant encouragement for him. The other author (DSM) expresses his unbounded love towards his daughters (Seema, Vandana and Anupama) who continuously and silently inspired him to write the book. Last but not least, our thanks are due to our wives Mrs. S. Mann and Mrs. K. Mukherjee for the patience and assistance rendered by them during various stages of the evolution of the book.

D. S. MANN
P. K. MUKHERJEE

Contents

Part One

RELATIVITY

Part One

RELATIVITY

1. Newtonian Mechanics

1.1. INTRODUCTION

Mechanics is the oldest of all the physical sciences. It is defined as that branch of physics which deals with the state of rest or motion of matter. Mechanics which deals with bodies at rest or in equilibrium under the action of forces or torques, is called *statics*. *Kinematics* describes abstract motion, i.e., the motion without considering its causes. Mechanics which relates motion to the forces or torques associated with it and also to the property of moving objects is known as *dynamics*.

The aim of mechanics is to analyse and explain various physical phenomena in terms of interactions exerted on each other by various physical systems. Once we understand the influences between the fundamental particles of matter, it becomes easy to find an explanation of all the physical phenomena of nature.

The three primary quantities with which we deal in mechanics are *space*, *time* and *mass*. Space or extension is derived by abstraction from extended objects. In physics the study of space means the measurement of length or the distance between two points.

Time is the mode of grouping sense impressions by the order in which the events are observed. Abstract time, as used in mechanics generally, is a parameter serving as the fundamental independent variable in terms of which the relative behaviour of all physical systems may be compared. Time may also be defined as the measure of duration of an event, though this is more properly referred to as a time interval. It is derived by abstraction from the motion of extended material objects in space and its measurement is based on the repetition of some apparently periodic event (periodic rotation of the earth).

Mass is the physical measure of the principal inertial property of a body, i.e., its resistance to change of motion. It is defined as the quantity of matter contained in a body and it is always conserved. It remains constant for speeds small compared with the speed of light but is variable for high velocities.

Based on the concept of space, time and mass, mechanics is classified further into following branches:

1. Classical or Newtonian mechanics
2. Relativistic or Einsteinian mechanics
3. Quantum mechanics
4. Relativistic Quantum mechanics

In classical mechanics space, time and mass are considered *absolute*, i.e., they are independent of the position or the motion of the object or the observer. Since it is based on Newton's laws of motion, it is also called Newtonian mechanics. Classical mechanics was built through the observations on macroscopic objects travelling at speeds small compared with the speed of light. Later on when the principles of classical mechanics were applied to microscopic objects having speeds approaching the speed of light, the results obtained did not fit into the experimental facts. This led to the formulation of a new mechanics which could hold good for all objects and for all speeds.

In 1905, a new scheme of mechanics, called relativistic mechanics, was developed by Einstein. This mechanics rejected the absolute nature of the basic quantities of space, time and mass and was applicable accurately to bodies travelling with all speeds. Here the space, time and mass which were considered absolute in classical mechanics were relative and varied with respect to the position and the motion of the object or the observer. Though the results of relativistic mechanics for high speed particles reduce to those of classical mechanics for slowly moving particles, it is convenient to use classical mechanics for small speeds as the equations of classical mechanics are much simpler than those of relativistic mechanics.

According to classical and relativistic mechanics the position and momentum of a particle can be measured simultaneously with any order of desired accuracy. In 1924, de Broglie, a French physicist, suggested that matter like radiation was dual in character. Each particle has a wave associated with it. This revolutionary concept led Heisenberg to formulate a principle known as

Heisenberg's uncertainty principle which not only shook the foundations of classical mechanics but brought about a radical change in our thinking concerning the laws of mechanics. According to this principle the position and momentum or the energy and the time of a particle cannot be measured simultaneously with any order of desired accuracy. If Δx is the range of values that might be found for the coordinate x of a particle, and Δp is the range in the simultaneous determination of the corresponding component of its momentum p, then

$$\Delta p \cdot \Delta x = \hbar = \frac{h}{2\pi} \tag{1.1a}$$

where h is the Planck constant. Similarly if ΔE and Δt are the uncertainties in the simultaneous determination of the energy and the time, then

$$\Delta E \cdot \Delta t = \frac{h}{2\pi} \tag{1.1b}$$

The wave aspect of matter is negligible for macroscopic objects but for microscopic particles like electrons, protons etc. the dual aspect (wave and particle) plays a dominant role. Under later circumstances our old concepts of space, time and mass have little meaning and this along with the experimental discoveries that the spectra of light emitted by gaseous elements consist of discrete lines and that atomic systems emit and absorb energy in a discontinuous fashion rather than a continuous stream, led to the re-examination of the foundations of the classical mechanics.

In 1925, therefore, Heisenberg and Schrodinger taking up the clue of matter waves developed independently two branches of mechanics. The one of Heisenberg is known as quantum mechanics and that of Schrodinger, wave mechanics. The phenomena on atomic scale which are inexplicable on the basis of classical mechanics can be accurately understood through the principles of quantum mechanics. For larger bodies and large distances where uncertainty principle is not applicable, the results of quantum mechanics reduce to those of classical mechanics.

Relativistic quantum mechanics (developed by Dirac in 1927) is a combination of the theory of relativity and quantum mechanics. It deals with microscopic objects travelling with speeds comparable with the speed of light. The results of relativistic quantum mechanics reduce to those of classical mechanics for large bodies having speeds negligible as compared to the speed of light.

The present century has seen a remarkable advancement in the various branches of physics. Many new discoveries and inventions have been made. In order to have a complete understanding of the old and new phenomena, it is essential for a student of physics to have a grasp over the various branches of mechanics. In this and subsequent chapters we shall deal with classical and relativistic mechanics and at suitable places wherever necessary, we shall make use of the results of quantum mechanics.

1.2. REST AND MOTION

When a body does not change its position with time it is said to be at rest. If the body changes its position with time it is said to be in a state of motion. Rest and motion may be absolute or relative. A body is said to be at absolute rest when it does not change its position with respect to a point which is stationary in space. If it changes its position with respect to a stationary point in space, it is said to be in absolute motion. As there is no point in space which is stationary (earth, moon, sun etc. all are moving), it is meaningless to talk about absolute rest and absolute motion.

In fact, both rest and motion are relative. A body is relatively at rest if it does not change its position with respect to its surroundings. On the other hand if it changes its position with respect to its environments, it is said to be in relative motion. For example a house on the earth is at rest with respect to the earth but it is in motion with respect to other heavenly objects like moon, sun, stars etc. This is because the earth is not only spinning about its own axis but is also revolving round the sun. Similarly a person sitting in a bus is at rest with respect to other passengers but he is in motion with respect to the outside objects like trees, buildings etc.

1.3. FRAME OF REFERENCE

In order to describe the state of rest or motion of an object we need a device or a system with respect to which the position of a body with time can be measured. *A frame of reference is a coordinate system attached to rigid body to describe the relative position of any particle in space.* It consists of a set of lines or surfaces which are used to define coordinates that in turn define positions, directions, velocities etc. of an object.

The simplest frame of reference is a cartesian coordinate system (Fig. 1.1) where the position vector $\vec{r}$ of a point from the origin is given by

$$\vec{r} = x\hat{i} + y\hat{j} + z\hat{k} \tag{1.2}$$

where $\hat{i}$, $\hat{j}$ and $\hat{k}$ are unit vectors along x, y and z axes respectively.

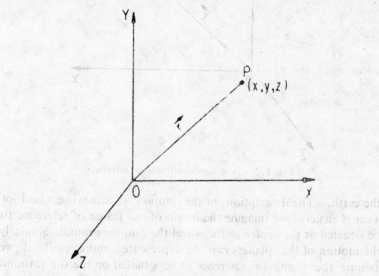

Fig. 1.1. A frame of reference.

The velocity and acceleration of the point are

$$\vec{V} = \frac{d\vec{r}}{dt} \text{ and } \vec{a} = \frac{d^2\vec{r}}{dt^2} \tag{1.3}$$

In physics a *point* is defined as the position of a physical phenomenon occurring in space. The time of occurrence of the physical phenomenon together with the point constitutes an *event*. An event is described by making use of a four coordinate system which is called space-time frame of reference (Fig. 1.2). The fourth coordinate is the time coordinate.

Motion of a body will look different for different frames of reference but we choose a frame of reference in which the motion of a body can be easily and accurately represented. For example the motion of an aeroplane with respect to another aeroplane may be quite complicated but its motion can he more conveniently described if the frame of reference is supposed to be situated at

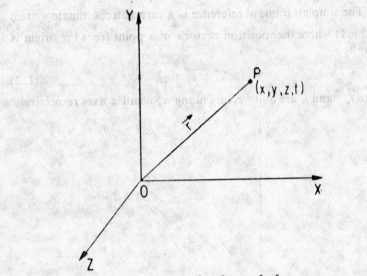

Fig. 1.2. A space-time frame of reference.

the earth. The description of the motion of a moving wheel of a car is easier if we imagine the origin of our frame of reference to be situated at the centre of the wheel than on the ground. Similarly the motion of the planets can be represented more easily if we assume the frame of reference to be situated on the sun (around which all planets move) than on the earth. Thus it is not necessary to have a frame of reference always on the earth. For a convenient description of motion it can be assumed to be situated even at a place which is not physically approachable.

1.4. NEWTON'S FIRST LAW OF MOTION

The discovery of the laws of motion was one of the revolutionary discoveries which changed the course of scientific history. For centuries the problem of motion and its causes had been one of the central themes of natural philosophy but it was left to the genius of Galileo and Issac Newton, respectively, to prepare the ground work and to lay the foundations of the laws of motion which are the bedrock of the principles of classical mechanics.

Before the time of Galileo it was thought that a body was in its 'natural state' when at rest and some influence or 'force' was essential to keep a body in motion. Galileo objected to this view and through his experiments on accelerated motion showed that no external force was necessary to maintain the velocity of a body. However, he asserted that a force was necessary to change the state of rest or motion of a body. This property of the body by virtue of which it was incapable to change by itself its state of rest or uniform motion in a straight line was termed as '*inertia*.' Mass was taken as the quantitative measure of inertia of a body.

This principle of Galileo (called the principle of inertia) was adopted and restated by Newton as his first law of motion which is stated as under:

"*Every body continues in its state of rest or uniform motion in a straight line unless it is compelled to change that state by some external agency called 'force'.*"

The first law of motion gives us the idea and definition of force. Thus force can be defined as some influence which produces or tends to produce, which stops or tends to stop the uniform motion of a body in a straight line. Therefore in the absence of forces or in the presence of forces whose resultant is zero, a body will persist in its state of rest or uniform motion in a straight line. So we conclude that if no force acts on a body, its acceleration is zero.

Mathematically Newton's first law of motion is written

$$\text{if} \quad \vec{F} = 0, \text{ then } \vec{a} = 0 \quad\quad (1.4)$$

where $\vec{F}$ is the force and $\vec{a}$ is the acceleration of a body.

1.5. NEWTON'S SECOND LAW OF MOTION

Newton's first law of motion gives the definition of force but it does not tell us how the force can be measured. The next logical step is to examine how a force can be measured and how a body behaves under the action of a resultant force. The answer to these questions is contained in Newton's second law of motion which states that *the rate of change of momentum is directly proportional to the impressed or the resultant force and takes place in the direction of force.*

Mathematically

$$\vec{F} \propto \frac{d\vec{p}}{dt} \quad\quad (1.5)$$

where $\vec{F}$ is the force and $\vec{p}$ is the momentum of the body on which the force acts.

Momentum of a body is the quantity of motion contained in it. It is measured by the product of mass and the velocity of the body; that is,

$$\vec{p} = m\vec{V} \tag{1.6}$$

where m is the mass and $\vec{V}$ is the velocity of the body.

Using Eq. (1.6) in Eq. (1.5), we get

$$\vec{F} \propto \frac{d}{dt}(m\vec{V})$$

$$\propto m \frac{d\vec{V}}{dt} \text{ (assuming the mass to be constant)}$$

or $\vec{F} = Km\vec{a}$ (1.7)

where K is a constant of proportionality and $\vec{a}\left(= \dfrac{d\vec{V}}{dt}\right)$ is the acceleration of the body. If we choose the unit of force to be such that $K = 1$, Eq. (1.7) can be written as

$$\vec{F} = m\vec{a} \tag{1.8}$$

Eq. (1.8) gives us the following informations:

1. Force can be measured if we know the mass and the acceleration of the body.
2. A body will move with accelerated motion in the presence of a force.
3. Mass of a body represents the force per unit acceleration.
4. If $\vec{F} = 0$, then $\vec{a} = 0$; that is, a body at rest will remain at rest and a body in motion will continue to move with a constant velocity. Hence Newton's first law is contained in the second law of motion.

Equation (1.8) also offers a method to measure and compare masses of two bodies. From Eq. (1.8)

$$m = \frac{\vec{F}}{\vec{a}} \tag{1.9}$$

Knowing $\vec{F}$ and $\vec{a}$, m can be measured. Further if a common force $\vec{F}$ acts on two bodies of masses m_1 and m_2, we may write

$$\vec{F} = m_1\vec{a}_1 = m_2\vec{a}_2$$

where $\vec{a}_1$ and $\vec{a}_2$ are respectively the accelerations of m_1 and m_2. Hence

$$\frac{m_1}{m_2} = \frac{\vec{a}_2}{\vec{a}_1} \qquad (1.10)$$

Equation (1.10) may be used to compare masses and if one of them is known the other can be calculated.

1.6. NEWTON'S THIRD LAW OF MOTION

Forces acting on a body are due to other bodies which constitute its environments. Any single force is but one aspect of mutual interaction between two bodies. Experiments have shown that when one body exerts a force on another, the later always exerts a force on the former. The magnitudes of these forces are found to be equal but opposite in directions. A single isolated force is therefore an impossibility.

The two forces involved in mutual interaction between two bodies are commonly known as 'action' and 'reaction'. Either force may be considered the 'action' and the other the 'reaction' to it.

The property of the forces mentioned above was incorporated by Newton in his third law of motion. In his own words, "To every action there is always opposed an equal reaction; or, the mutual actions of two bodies upon each other are always equal, and directed to contrary parts."

Thus if a body A exerts a force $\vec{F}_{AB}$ on a body B, the body B exerts an equal but oppositely directed force $\vec{F}_{BA}$ on body A; that is,

$$\vec{F}_{AB} = -\vec{F}_{BA} \qquad (1.11)$$

Forces of action and reaction occur in pairs and act on different bodies. They act along the line joining the two bodies. If action and reaction act on the same body, we shall never get an accelerated motion as the resultant force on every body will then always be zero.

As we have seen that mass may be measured by Eq. (1.10), but this involves the idea of a common (equal) force. How can we apply such a force on two bodies unless we know their masses already? This possibility of applying equal forces is provided by Newton's third law of motion. Since we now have a means of measuring mass Newton's second law of motion becomes meaningful. As first law is contained in the second law, the two laws become useful only with the aid of the third law.

Further it can be shown that Newton's third law of motion is contained in the second law of motion and that the second law is the real law of motion. Consider a wooden block on a table. The block exerts an action on the table and the table exerts an equal and opposite reaction on the block. If action and reaction are not equal, the block or the table will change its state without the help of an external force. But this is forbidden by the first law of motion. Therefore, action has got to be equal and opposite to the reaction. Thus the third law is contained in the first law but the first law is only a special case of the second law of motion. So the third law of motion is also contained in the second law, though the first and the second laws are meaningful only in the presence of the third law of motion.

1.7. APPLICATIONS AND LIMITATIONS

The applications of the laws of motion are so many that they are considered as the heart and soul of classical mechanics. From simple cases of the rectilinear motion, the motions of pendulums, oscillators, oscillating masses etc., they can be applied to explain the motion of planets and also to calculate the perturbations on a planet due to other planets. Some important cosequences of Newton's laws of motion are the laws of conservation of mass, energy and momentum.

Newton's laws of motion apply strictly to infinitesimal particles, but may be extended to rigid bodies by the assumption that these bodies may be treated as collection of particles whose masses are concentrated at a single point. Laws of motion cannot be applied to problems as such. We have to conceive a model of the situation to which the laws can be applied. In order to apply the laws, the body or the bodies under consideration are firstly reduced to a point and then the environments are replaced by suitable forces acting at that point.

Newton's laws of motion have certain limitations. Firstly they hold only when space, time and mass are considered invariant. The second law and the first law do not hold good for an observer who is accelerated. The third law is valid when action and reaction are measured at the same time. But the forces have a finite propagation velocity as the reaction produced on a particle due to the action of another particle reaches the later after a finite interval of time. Thus at any instant, action may not be equal to reaction. For example a charged particle moving towards the atom, causes electric polarisation in it and hence an electric field will be generated in the atom. The effect of this electric field will reach the charged particle after some finite time. Therefore, we can say that the third law of motion is not strictly true.

Physics deals with the nature by reducing it to simplified models to which the laws of physics can be applied. Every law makes certain basic assumptions applicable only to a definite model. The law necessarily fails when we can not represent the physical situation by a presupposed model satisfying the basic assumptions of the law. Thus it is found that Newton's laws fail when applied to motion of particles in an atom, and also when applied to bodies having speeds comparable to the speed of light. In both these situations the basic assumptions made in the structure of Newton's laws are violated.

1.8. INERTIAL FRAMES OF REFERENCE

A frame of reference in which Newton's laws of motion are valid, is called an inertial frame. Such a frame is unaccelerated and non-rotating. In an inertial frame a body will continue in its state of rest or uniform rectilinear motion in the absence of an external force. Reference frames moving with uniform velocity with respect to each other and with respect to fixed stars are the examples of inertial frames. Also a reference frame attached to the earth can be considered to be an inertial frame for most practical purposes. The word inertial has been derived from Newton's first law of motion which is often called the law of inertia.

It can be shown by experiments that all inertial frames are equivalent for the measurement of physical phenomena. Physical quantities may have different values for different observers in different inertial frames but the basic laws of physics (the relationship between the measured quantities) will always remain the same for all observers. For example, let two observers in different inertial frames measure the momenta of two particles before and after the collision. It is found that they will obtain different numerical values both for the momenta of individual particles and for the total momentum of two particles. Each observer, however, will note that the total momentum is the same before and after the collision in their respective frames. In other words the law of conservation of momentum is the same in both the frames of reference.

Although the physical laws remain invariant in all inertial frames, the measured values of the physical quantities may be different in different frames. It is, therefore, important for the students to realise what their frame of reference is in a particular problem.

1.9. FRAMES OF REFERENCE IN UNIFORM RELATIVE MOTION

Let S be an inertial frame represented by cartesian coordinate system (O, XYZ). S', another frame of reference represented by cartesian coordinate system $(O', X'Y'Z')$, is moving with a uniform relative velocity $\vec{v}$ with respect to S (Fig. 1.3). At time $t = 0$, the origins of the two systems, O and O', coincide with each other. After any time t, $OO' = \vec{vt}$. Let the position vectors of any particle P in the S and S' systems be $\vec{r}$ and $\vec{r}'$ respectively.

By addition of vectors

$$\vec{r} = \vec{r}' + \vec{v}t. \tag{1.12}$$

Differentiating both sides with respect to time (assuming $t = t'$).

$$\frac{d\vec{r}}{dt} = \frac{d\vec{r}'}{dt'} + \vec{v}$$

or $$\vec{V} = \vec{V}' + \vec{v} \tag{1.13}$$

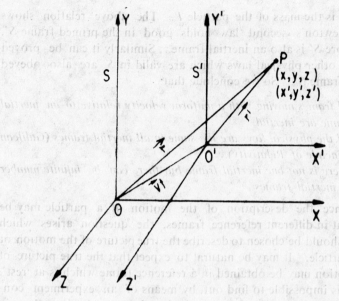

Fig. 1.3. Frames of reference in uniform relative motion.
S' is moving w.r.t. S.

where $\vec{V}$ and $\vec{V'}$ are observed velocities of particle P in S and S' systems respectively.

Equation (1.13) transforms the velocity of the particle from unprimed frame (S) to primed frame (S') and is known as the Galilean law of addition of velocities.

Differentiating Eq. (1.13) with respect to time, we get

$$\frac{d\vec{V}}{dt} = \frac{d\vec{V'}}{dt'} + 0$$

or
$$\vec{a} = \vec{a'} \tag{1.14}$$

where $\vec{a}$ and $\vec{a'}$ are, respectively, the accelerations of the particle in S and S'. Since S is an inertial frame, Newton's laws are valid in it. Hence we can write

$$\vec{F} = m\vec{a} \qquad \text{(II law)}$$

or
$$\vec{F} = m\vec{a'} \qquad (\because \vec{a} = \vec{a'})$$

Here m is the mass of the particle P. The above relation shows that Newton's second law holds good in the primed frame S'. Therefore S' is also an inertial frame. Similarly it can be proved that all other physical laws which are valid in S are also obeyed in the frame S'. So we conclude that:

1. *All frames moving with a uniform velocity relative to an inertial frame are inertial.*
2. *All the physical laws are the same in all inertial frames (Galilean principle of Relativity).*
3. *There is not one inertial frame but there can be infinite number of inertial frames.*

Since the description of the motion of a particle may be different in different reference frames, the question arises which frame should be chosen to describe the true picture of the motion of the particle. It may be natural to expect that the true picture of the motion may be obtained in a reference frame which is at 'rest'. But it is impossible to find out, by means of an experiment conducted in a frame of reference, whether the frame is at 'rest' or in uniform 'motion'. What can be detected is the motion of a frame relative to another frame. For example, we suppose an observer sitting inside a train moving with a uniform velocity. Let him throw a stone vertically upwards. He finds that the stone falls along the same vertical path (Fig. 1.4a). But for a stationary observer on the ground, the stone will appear to be falling along a parabolic path (Fig. 1.4b). If the observer on the ground per-

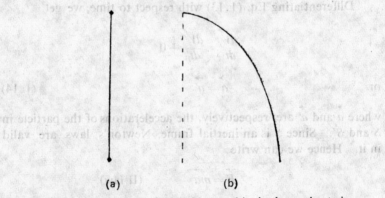

(a)   (b)

Fig. 1.4.   Path as seen by an observer: (a)   in the moving train;
(b)   on the ground.

forms the same experiment, the result will be vice versa. Therefore a particle may appear to be moving along a straight line when observed in one frame of reference, while it may appear to be moving along a parabola when observed from another frame of reference. As there is no means of deciding which reference frame is at rest, there is also no means of deciding which of the two paths is the 'real' path. What we can decide is that the motion of the particle is simpler in one frame than that in the other reference frame. Thus the choice of the reference frame is made on the basis of the simplicity of the motion of the particle in that frame.

1.10. GALILEAN TRANSFORMATIONS

At any instant the coordinates of a particle are different in different reference frames. But the coordinates can be transformed from one reference frame to another with the help of mathematical equations. *The transformation equations for the inertial frames are known as Galilean transformations.*

Let S and S' be two space-time frames of reference formed by two cartesian coordinates systems (O, XYZ) and $(O', X'Y'Z')$ respectively (Fig. 1.5). Their x-axes coincide and S' is moving

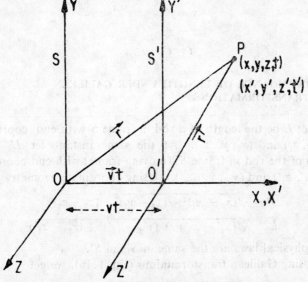

Fig. 1.5. Space-time reference frames in relative motion.

with a uniform velocity $\vec{v}$ relative to S along x-axis. At $t = 0$, the origins O and O' coincide. P is an event to be observed either from S or S'. We further assume that space and time are absolute, i.e., they are independent of any particular frame of reference ($t = t'$ and $L = L'$).

Under such circumstances, the position and time coordinates of P are related by the equation [using Eq. (1.12)].

$$\vec{r}' = \vec{r} - \vec{v}t$$
$$t' = t \tag{1.15}$$

In the component form Eq. (1.15) is written as

$$x' = x - vt$$
$$y' = y$$
$$z' = z \tag{1.16}$$
$$\text{and} \qquad t' = t$$

Eq. (1.16) expresses the transformation of coordinates from one inertial frame (S) to another (S') and hence the relations of Eq. (1.16) are called Galilean transformations.

The inverse Galilean transformations are expressed as

$$x = x' + vt$$
$$y = y'$$
$$z = z' \tag{1.17}$$
$$t = t'$$

1.11. INVARIANCE OF LENGTH UNDER GALILEAN TRANSFORMATIONS

Let L be the length of a rod in frame S with end coordinates (x_1, y_1, z_1) and (x_2, y_2, z_2). At the same instant let L' be the length of the rod in frame S' (moving frame) with end coordinates (x_1', y_1', z_1') and (x_2', y_2', z_2'). Using coordinate geometry

$$L = \sqrt{(x_2 - x_1)^2 + (y_2 - y_1)^2 + (z_2 - z_1)^2} \tag{1.18}$$

$$\text{and} \qquad L' = \sqrt{(x_2' - x_1')^2 + (y'_2 - y_1')^2 + (z_2' - z_1')^2} \tag{1.19}$$

since physical laws are the same in S and S'.

Using Galilean transformtaions (Eq. 1.16), we get

$$x_2' - x_1' = x_2 - x_1$$

$$y_2' - y_1' = y_2 - y_1 \qquad (1.20)$$

and $\qquad\qquad z_2' - z_1' = z_2 - z_1$

Substituting Eq. (1.20) in Eqs. (1.18) and (1.19), we get

$$L = L'$$

Hence length is invariant under Galilean transformations.

1.12. INVARIANCE OF NEWTON'S LAWS OF MOTION UNDER GALILEAN TRANSFORMATIONS

Let S and S' be two inertial frames. S' is moving with a uniform velocity relative to S. A particle P of mass m has accelerations $\vec{a}$ and $\vec{a}'$ in S and S' frames respectively. Since S and S' are inertial frames, Newton's laws of motion are valid in both of them. Hence according to Newton's second law

$$\vec{F} = m\vec{a}$$

and $\qquad\qquad \vec{F}' = m\vec{a}' \qquad (1.21)$

where $\vec{F}$ and $\vec{F}'$ are the forces acting on P in frames S and S' respectively.

Now differentiating Eq. (1.15) twice with respect to time, we get

$$\frac{d^2\vec{r}'}{dt'^2} = \frac{d^2\vec{r}}{dt^2} \qquad (\because \vec{v} = \text{const})$$

or $\qquad\qquad \vec{a}' = \vec{a} \qquad (1.22)$

Using Eq. (1.22) in Eq. (1.21), we have

$$\vec{F} = \vec{F}'$$

This means that the force $(\vec{F} = m\vec{a})$ has the same magnitude and direction in the frames S and S' which are moving with a uniform relative motion. Hence Newton's second law is invariant under Galilean transformations. We have already proved that Newton's first law $(\vec{F} = 0, \vec{a} = 0)$ and the third law (involving forces) are contained in the second law. So we conclude that Newton's laws of motion remain unaffected under Galilean transformations.

1.13. CONSERVATION OF LINEAR MOMENTUM

The law of conservation of momentum states that the total linear momentum of a system remains constant in the absence of external forces. As the linear momentum is a vector quantity, the total linear momentum of a system of n particles in a given reference frame is given by the vector sum of the momenta of the individual particles in the same frame; that is,

$$\vec{P} = \vec{p_1} + \vec{p_2} + \vec{p_3} + \ldots \vec{p_n} \tag{1.23}$$

where $\vec{P}$ is the total linear momentum and $\vec{p_1}$, $\vec{p_2}$ etc. are the momenta of individual particles.

Using Newton's second law of motion, the total force acting on the system is given by

$$\vec{F} = \frac{d\vec{P}}{dt} = \frac{d}{dt}\left(\vec{p_1} + \vec{p_2} + \vec{p_3} + \ldots \vec{p_n}\right) \tag{1.24}$$

But the total force $\vec{F}$ acting on the system is the sum of external and internal forces; that is,

$$\vec{F} = \vec{F}_{ext} + \vec{F}_{int} \tag{1.25}$$

where $\vec{F}_{ext}$ is the external force and $\vec{F}_{int}$ (the internal force) is due to the mutual interaction of the particles on one another. $\vec{F}_{int}$ is given by Newton's third law of motion which states that internal forces given by action and reaction exist in pairs. Since action and reaction are equal in magnitude and opposite in direction, they annul each other. Therefore

$$\vec{F}_{int} = 0 \tag{1.26}$$

Using Eqs. (1.25) and (1.26) in Eq. (1.24), we get

$$\vec{F}_{ext} = \frac{d\vec{P}}{dt} = \frac{d}{dt}\left(\vec{p_1} + \vec{p_2} + \ldots \vec{p_n}\right) \tag{1.27}$$

When the external forces acting on the system are zero, $\vec{F}_{ext} = 0$ and Eq. (1.27) gives

$$\vec{P} = \vec{p_1} + \vec{p_2} + \vec{p_3} + \ldots \vec{p_n} = \text{constant} \tag{1.28}$$

Equation (1.28) is the mathematical statement of conservation of linear momentum.

Let us consider a system of two particles of masses m_1 and m_2 moving with velocities $\vec{U_1}$ and $\vec{U_2}$ respectively. If they are allowed to collide, their velocities become $\vec{V_1}$ and $\vec{V_2}$ respectively. Hence their total linear momentum before collision is $m_1\vec{U_1} + m_2\vec{U_2}$ and after the collision the total momentum becomes $m_1\vec{V_1} + m_2\vec{V_2}$. Applying the law of conservation of momentum we can write

$$m_1\vec{U_1} + m_2\vec{U_2} = m_1\vec{V_1} + m_2\vec{V_2} \qquad (1.29)$$

Eq. (1.29) tells us that the momentum of the individual particles may change but the total momentum remains constant.

We have seen that Newton's third law of motion has played a key role in the above deduction of the law of conservation of linear momentum. The third law has been used to justify the assumption that the sum of internal forces $(\vec{F}_{int})$ acting on the system is zero. However, we have noted that this law has a limitation due to finite propagation velocity of the forces of action and reaction (particularly in case of atomic particles) and, at any instant, action may not be equal to reaction. Thus, there will be a resultant internal force acting on the system $(\vec{F}_{int} \neq 0)$. Therefore we give below a second method to prove the validity of the law of conservation of momentum.

Second Method. This method is more general and is based on Galilean invariance and the law of conservation of energy.

Let us consider the collision of two particles m_1 and m_2 with initial and final velocities $\vec{U_1}, \vec{U_2}$ and $\vec{V_1}, \vec{V_2}$ respectively in a frame S. If we assume the collision to be inelastic (a general case), then the law of conservation of energy in the frame S is given by

$$\tfrac{1}{2}m_1 U_1^2 + \tfrac{1}{2}m_2 U_2^2 = \tfrac{1}{2}m_1 V_1^2 + \tfrac{1}{2}m_2 V_2^2 + E \qquad (1.30)$$

where E is the energy lost during the collision in the form of heat, light etc. Here m_1 and m_2 have been assumed to be constant.

According to the principle of invariance of Galileo, the laws of physics remain invariant under Galilean transformations. Hence, assuming the energy lost during the collision to be the same (an experimental fact), the law of conservation of energy in a frame S'

moving with a uniform velocity $\vec{V}$ with respect to S is written as

$$\tfrac{1}{2}m_1{U_1'}^2 + \tfrac{1}{2}m_2{U_2'}^2 = \tfrac{1}{2}m_1{V_1'}^2 + \tfrac{1}{2}m_2{V_2'}^2 + E \qquad (1.31)$$

where $\vec{U'}_1$, $\vec{U'}_2$, and $\vec{V'}_1$, $\vec{V'}_2$ are, respectively, initial and the final velocities of the particles m_1 and m_2 in the frame S'.

Using Galilean law of addition of velocities (Eq. 1.13), we have

$$\vec{U_1'} = \vec{U_1} - \vec{V}, \quad \vec{U_2'} = \vec{U_2} - \vec{V}$$

$$\vec{V_1'} = \vec{V_1} - \vec{V}, \vec{V_2'} = \vec{V_2} - V \qquad (1.32)$$

Substituting Eq. (1.32) in Eq. (1.31), we get

$$\tfrac{1}{2}m_1\,(\vec{U_1} - \vec{V})^2 + \tfrac{1}{2}m_2\,(\vec{U_2} - \vec{V})^2 = \tfrac{1}{2}m_1(\vec{V_1} - \vec{V})^2$$
$$+ \tfrac{1}{2}m_2(\vec{V_2} - \vec{V})_2 + E$$

or $\tfrac{1}{2}m_1\,(\vec{U_1} - \vec{V}) \cdot (\vec{U_1} - \vec{V}) + \tfrac{1}{2}m_2(\vec{U_2} - \vec{V}) \cdot (\vec{U_2} - \vec{V})$

$$= \tfrac{1}{2}\,m_1\,(\vec{V_1} - \vec{V}) \cdot (\vec{V_1} - \vec{V}) + \tfrac{1}{2}m_2\,(\vec{V_2} - \vec{V}) \cdot (\vec{V_2} - \vec{V}) + E$$

or $\tfrac{1}{2}m_1\,[U_1^2 + V^2 - 2\vec{U_1} \cdot \vec{V}] + \tfrac{1}{2}m_2\,[U_2^2 + V^2 - 2\vec{U_2} \cdot \vec{V}]$

$$= \tfrac{1}{2}\,m_1\,[V_1^2 + V_1^2 - 2\vec{V_1} \cdot \vec{V}]$$
$$+ \tfrac{1}{2}\,m_2\,[V_2^2 + V^2 - 2\vec{V_2} \cdot \vec{V}] + E$$

or $\tfrac{1}{2}\,m_1U_1^2 - m_1\vec{U_1} \cdot \vec{V} + \tfrac{1}{2}\,m_2U_2^2 - m_2\,\vec{U_2} \cdot \vec{V}$

$$= \tfrac{1}{2}\,m_1V_1^2 - m_1\vec{V_1} \cdot \vec{V} + \tfrac{1}{2}m_2V_2^2 - m_2\vec{V_2} \cdot \vec{V} + E$$

or $\tfrac{1}{2}\,m_1U_1^2 + \tfrac{1}{2}\,m_2U_2^2 + m_1\,(\vec{V_1} \cdot \vec{V} - \vec{U_1} \cdot \vec{V})$

$$= \tfrac{1}{2}m_1V_1^2 + \tfrac{1}{2}\,m_2V_2^2 + m_2\,(\vec{U_2} \cdot \vec{V} - \vec{V_2} \cdot \vec{V}) + E \qquad (1.33)$$

Subtracting Eq. (1.30) from Eq. (1.33), we get

$$m_1\,(\vec{V_1} - \vec{U_1}) \cdot \vec{V} = m_2(\vec{U_2} - \vec{V_2}) \cdot \vec{V}$$

Hence $m_1\vec{U_1} + m_2\vec{U_2} = m_1\vec{V_1} + m_2\vec{V_2} \qquad (1.34)$

Equation (1.34) proves the law of conservation of linear momentum.

The law of conservation of momentum is valid even in the field of atomic and nuclear physics where Newtonian mechanics fails. Hence the principle of conservation of momentum is more exact, fundamental and universal than the Newtonian principles. In fact the laws of conservation of mass, momentum and energy are the basic laws of nature. These laws are their own proofs and do not need deductions from any other consideration.

1.14. LINEAR SYMMETRY

Nature loves symmetry. If we observe the objects of nature, we shall always note a kind of symmetry in them. Symmetry is also fascinating to human mind and perhaps the most symmetrical object imaginable is a sphere. Heavenly bodies like the sun, the earth, the moon etc are all spherical in shape and form.

In an ordinary sense an object is said to be symmetrical if one side of it appears to be the same as the other side. According to Professor Hermann Weyl, an object is said to be symmetrical if it appears to be the same before and after a certain operation. For example, take a vase which is left-and-right symmetrical. If we turn it through 180° about a vertical axis, it would appear to be the same. *Symmetry in physics involves a more general view of the definition of symmetry by Weyl. This is the symmetry of physical laws.* In physics there are many operations under which physical phenomena remain invariant. It has been found that the symmetry exists in the basic laws which govern the operation of the physical world. In what follows we confine ourselves to the symmetry that is found to exist in linear motion.

Linear symmetry in physics meant that the laws of physics remain invariant when we operate them in systems moving with uniform velocity with respect to each other. Physical equations, when displaced in space, reproduce themselves and remain unchanged after displacement. Therefore we say that the laws of physics are symmetrical for translational displacements in space; that is, the laws do not change when we make a translation of our coordinates.

Let us prove the above mentioned property of symmetry in physics by operating upon Newton's second law of motion. A and B are two observers stationed in the frames $S(XYZ, O)$ and S' $(X'Y'Z', O')$ respectively (Fig. 1.6). S' is moving with a uniform velocity $\vec{V}$ with respect to S along x-axis. O, the origin of observer A is fixed with respect to O', the origin of observer B. At $t = 0$, O and O' coincide.

Relation between the coordinates of a point P, in the two frames, are given by Galilean transformations

$$x' = x - Vt,$$

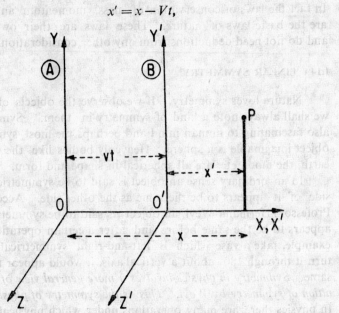

Fig. 1.6.   Two parallel coordinate systems.

$$y' = y$$
$$z' = z,$$ (1.35)

and

$$t' = t$$

In the frames S and S', respectively, Newton's second law is written as

$$\vec{F} = m\frac{d^2\vec{r}}{dt^2}$$ (1.36)

$$\vec{F'} = m\frac{d^2\vec{r'}}{dt'^2}$$ (1.37)

where $\vec{F}$ and $\vec{F'}$ are, the forces acting on the particle P of mass m in the frames S and S' respectively.

In the component form Eqs. (1.36) and (1.37) are expressed as

$$F_x = m\frac{d^2x}{dt^2}, \quad F_y = m\frac{d^2y}{dt^2}, \quad F_z = m\frac{d^2z}{dt^2}$$ (1.38)

and $$F_x' = m\frac{d^2x'}{dt'^2}, \quad F_y' = m\frac{d^2y'}{dt'^2}, F_z' = m\frac{d^2z'}{dt'^2} \qquad (1.39)$$

Using Eq. (1.35), we write

$$\frac{d^2x'}{dt'^2} = \frac{d^2x}{dt^2}, \frac{d^2y'}{dt'^2} = \frac{d^2y}{dt^2}, \frac{d^2z'}{dt'^2} = \frac{d^2z}{dt^2} \qquad (1.40)$$

Substituting Eq. (1.40) in Eq. (1.39), we get

$$F_x' = m\frac{d^2x}{dt^2}, \quad F_y' = m\frac{d^2y}{dt^2}, \quad F_z' = m\frac{d^2z}{dt^2} \qquad (1.41)$$

Comparing Eqs. (1.38) and (1.41), we conclude

$$F_x' = F_x, \quad F_y' = F_y \text{ and } F_z' = F_z$$

Hence $$\vec{F'} = \vec{F}$$

This means that Newton's second law remains invariant for an observer B in a frame S' which is moving with a uniform velocity $\vec{V}$ with respect to S, the frame of another observer A. Thus we see that the physical equations remain invariant during linear displacements. For example, a machine operated by an operator A at one place will behave in the same manner as a similar machine, at another place, does when it is operated by an another operator B. This is because the laws, governing the similar machines, are the same for the two equally efficient operators. This is known as the symmetry of translation in space in physics.

QUESTIONS AND PROBLEMS

1.1. What do you understand by (a) space, time and mass (b) a frame of reference and (c) an inertial frame of reference?

1.2. State and discuss Newton's laws of motion. Show that Newton's second law is the real law of motion. Mention some of the applications and limitations of Newton's laws.

1.3. Define and discuss Galilean transformations. Prove that Newton's laws of motion are invariant under Galilean transformations.

1.4. State and prove the law of conservation of linear momentum.

1.5. What do you understand by symmetry in physics? Discuss the linear symmetry of physical laws.

1.6. A cricket ball is dropped from an aeroplane flying at a constant horizontal velocity. Describe its motion relative to (a) pilot and (b) an observer on the earth.

[(a) Straight line, (b) Parabola]

1.7. Can a person sitting in a car detect the state of uniform motion of his car with respect to the ground by looking at the speedometer? If not why? Does this violate the principle of relativity?

1.8. When set up in a laboratory, the length, time period and the mass of the bob of a simple pendulum are 1 metre, 2 seconds and 0.1 kg respectively. If it is set up in a train moving with a uniform horizontal velocity, what will be the values of the length, time period and mass of the bob of the pendulum when the velocity of the train is (a) 20 km/hr and (b) 40 km/hr?

$$\begin{bmatrix} \text{(a)} & 1 \text{ m, 2 sec, 0.1 kg} \\ \text{(b)} & 1 \text{ m, 2 sec, 0.1 kg} \end{bmatrix}$$

1.9. An object is moving with an acceleration of 10 metres/sec² when observed by a person at rest. What will be its acceleration when it is observed by a person sitting in a car moving with a uniform horizontal velocity of 25 km/hr? If the mass of the object is 2 kg, what is the force acting on it as measured by (a) an observer at rest on the ground and (b) an observer in the moving car?

$$\begin{bmatrix} \text{Acceleration} = 10 \text{ m/sec}^2, \\ \text{Force} = 20 \text{ Newtons in both frames} \end{bmatrix}$$

1.10. (a) Discuss Galilean law of addition of velocities.

(b) The position vector of a particle in an inertial frame S is

$$\vec{r} = (8t^2 - 4t)\,\hat{i} + (-6t^3)\,\hat{j} + 5\hat{k}$$

Determine the constant relative velocity of reference frame S' with respect to S if the position vector of the same particle in S' is

$$\vec{r'} = (8t^2 + 3t)\,\hat{i} + (-6t^3)\,\hat{j} + 5\hat{k}$$

Also show that the acceleration of the particle is the same in both frames of reference.

[(b) $-7\hat{i}$]

1.11. A 40 kg man standing on a surface of negligible friction kicks forward a 10 gm stone lying at his feet so that it acquires a speed of 4 metres/sec. What velocity does the man acquire as a result?

[0.1 cm/sec]

1.12. A shell of 5 kg mass moving with a velocity of 20 km/sec explodes above the atmosphere into two pieces which move in the same direction with 25 km/sec and 15 km/sec velocities. Determine the mass of each piece and also the energy set free by explosion.

[Hint. Energy set free = Final energy − Initial energy]

$$\left[\begin{array}{l}\text{Mass of each piece} = 2.5 \text{ kg} \\ \text{Energy set free} = 62.5 \times 10^6 \text{ Joules}\end{array}\right]$$

1.13. A bomb in flight explodes into two fragments when its velocity is $5\hat{i} + 4\hat{j}$. If the smaller mass M flies with a velocity $10\hat{i} + 100\,\hat{j}$, determine the velocity of the larger fragment whose mass is $3M$.

$$[\tfrac{1}{3}(10\hat{i} - 84\hat{j})]$$

2. Einsteinian Relativity and the Lorentz Transformations

2.1. INTRODUCTION

The concept of an absolute space and time was conceived by Newton. However, with the advent of relativity theory proposed by Einstein these concepts had to be fundamentally altered. It is now established that space and time are not absolute but are retative entities. The theory of relativity incorporates many novel ideas which tease the imagination. These ideas have considerable importance in discussing most atomic phenomena. In short, the theory of relativity examines how the measurements of length, mass, and time depend on the observer. It should be remarked that the consequences of relativity theory manifest only at very high velocities approaching the velocity of light. These velocities are called *relativistic* velocities. At non-relativistic velocities ($v \ll c$) the results obtained from the relativity theory reduce to the classical results of Newtonian theory.

2.2. THE ETHER HYPOTHESIS

It will be recalled that Newton chose the frame of the so-called fixed stars as the reference frame to enunciate his laws of motion. However, even in this frame Newton's laws are only valid approximately because this frame is not strictly the one which is *absolutely* at rest. The search for something more fixed than stars went on in the hope that a fixed reference frame would be discovered relative to which Newton's laws would hold exactly. It is in connection with this search that the existence of a hypothetical medium called ether was proposed. It was thought that ether was absolutely at rest and so a reference frame attached to ether might serve as a fixed reference frame. The necessity for assuming the existence of ether stemmed from the fact that every wave motion must need a medium for its propagation. After all, there cannot be sound waves in a vacuum, or ocean waves without an ocean. So, it was argued that light waves must also involve waving of something for their propagation even through free space. This medium was given the name luminiferous ether.

However, difficulties arose regarding nature of ether. Since light passes through material media as also through vacuum, ether must permeate all space and fill all matter. Hence ether must be regarded as highly rarefied form of medium. This property of ether is also necessary in order to explain the apparently unrestricted motion of heavenly bodies in space. But transverse nature of light as borne out by Maxwell's theory and experimental observations on polarized light, demands on the other hand that ether should be regarded as solid. This is essential since only solids can stand the shear forces associated with transverse waves.

The envisaged properties of ether are obviously self contradictory. How can a medium be rarefied and at the same time act like a highly dense solid? But so fascinated were the physicists at that time with the concept of ether that they even tried to do away with this anomaly by arguing that although the ratio of elasticity to density of ether may be large, yet individually both of these quantities are very small. Lord Kelvin estimated that the density of ether is of the order of 5×10^{-15} kg/m^3. Taking the velocity of light in vacuum as 3×10^8 m/sec, the rigidity coefficient η works out to be $\dfrac{\eta}{5 \times 10^{-15}} = c^2 = 9 \times 10^{16}$ or $\eta = 4.5 \times 10^2$ N/m^2 which indeed is a small quantity. Ether with this low rigidity value is expected to offer only vanishingly small resistance to a solid moving through it.

Our foregoing discussion indicates that however controversial the issue of ether might have been, physicists felt that it might be just the thing to which to attach a Newtonian coordinate system. It was conceived that Newton's laws would hold exactly for an observer moving with uniform velocity relative to the ether.

Now, if there is an ether which remains stationary and if the earth travels through it, then there must be an ether wind, just as a person riding on a bicycle through still air feels an air blowing in his face. The earth must move through the ether with its orbital velocity around the sun which is 3×10^4 m/sec (18.5 mi/sec). If the sun is also in motion then the drift of the earth might become even greater than its orbital velocity at some seasons. However, the earth's orbital velocity is only about 10^{-4} times the velocity of light. It is thus evident that a very sensitive instrument is required to test the existence of ether wind.

2.3. THE MICHELSON-MORLEY EXPERIMENT

A device of sufficient sensitivity was used by Albert A. Michelson and Edward W. Morley in 1887 to find the relative motion of earth with respect to ether. But the experiment gave a null result, i.e., no ether wind could be detected. The apparatus employed was essentially a Michelson interferometer, sketched in Fig. 2.1, floating on a pool of mercury.

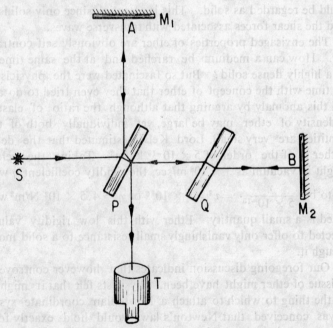

Fig. 2.1. The Michelson interferometer.

A beam of light from a monochromatic source S strikes a half-silvered glass plate P inclined at an angle of 45° to the initial direction of beam. Half the intensity of the incident light falling on P is reflected to the mirror M_1 while the other half is transmitted through P. The transmitted light passes through a compensating glass plate Q (which has the same thickness and inclination as P) and falls normally on the mirror M_2. The reflected and the transmitted rays are returned by the mirrors M_1 and M_2 respectively to the glass plate P where they are again partly reflected and partly transmitted. The transmitted part of the light from M_1 joins the reflected part of the light from M_2 and the combined light passes into the telescope T. The glass plate Q makes the optical path lengths of the reflected and the transmitted beams equal because the two light beams are made to traverse equal number of glass thicknesses.

If the earth were stationary the two path lengths would have become exactly equal making the path difference zero. However, in actual practice the apparatus is moving with the earth through the stationary ether. So, the paths of the two beams become unequal and depending upon whether this path difference is an even or odd multiple of half wavelength, a bright or a dark line will be observed. In fact what the observer sees through the telescope is multiplicity of lines called interference fringes like the teeth of a comb.

In order to calculate the path difference between the reflected and the transmitted beams, it is convenient to calculate the times taken by the two beams for their journeys to the respective mirrors and back to the plate P. Let the direction of motion of earth coincide with the direction of propagation of the initial beam of light. Due to the motion of earth the positions of the various components of the interferometer change and the reflections at the mirrors take place at A' and B' (Fig. 2.2). The new positions of the glass plate P, the mirrors and the telescope are as shown by the dotted lines. The plate Q is however not shown for convenience.

Let t_1 be the time taken by the reflected light in going from P to A' and returning from A' to P'; and t_2 the corresponding time taken by the transmitted light in traversing the distance from P to B' and from B' to P'. These times can be evaluated as follows:

Let c be the velocity of light through ether and v the velocity of earth carrying the apparatus with it. Let l be the dis-

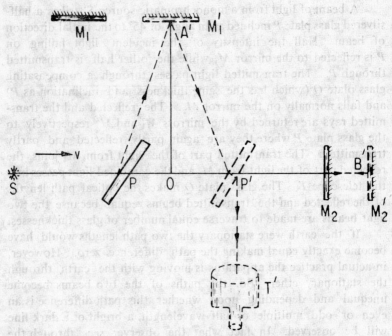

Fig. 2.2. The new position of the interferometer.

tance PA or PB. If the time required by the reflected light for its journey from P to A' or from A' to P' be τ, then from Fig. 2.2 we can write

$$c^2\tau^2 = v^2\tau^2 + l^2 \qquad (2.1)$$

Simplifying

$$\tau = \frac{l}{c(1 - v^2/c^2)^{1/2}} \qquad (2.2)$$

We now make use of the binomial expansion $(1 \pm x)^n \simeq 1 \pm nx$ which is true for $x \ll 1$. This expansion is valid in the present case because for the earth's orbital velocity, $(v/c)^2$ is of the order of 10^{-8}. Hence we get

$$\tau = \frac{l}{c}\left(1 + \tfrac{1}{2}\frac{v^2}{c^2}\right) \qquad (2.3)$$

The total time taken by the reflected ray in traversing the whole path $PA'P'$ is therefore given by

$$t_1 = 2\tau = \frac{2l}{c}\left(1 + \tfrac{1}{2}\frac{v^2}{c^2}\right) \qquad (2.4)$$

Now, to obtain t_2 we must calculate the times taken by the transmitted light for its forward trip from P to B' and the return trip from B' to P'. If the respective times for these two trips be t and t', we have

$$ct = l + vt \tag{2.5}$$

$$ct' = l - vt' \tag{2.6}$$

Eq. (2.5) indicates that the transmitted light in its forward journey from P to B' takes a path which is greater than l. This is because in the time interval t, the mirror M_2 also moves through a distance vt. A similar argument holds for Eq. (2.6). The total time taken by the light for the whole journey $PB'P'$ is

$$t_2 = t + t' = \frac{l}{c - v} + \frac{l}{c + v} \tag{2.7}$$

Eq. (2.7) gives upon simplification

$$t_2 = \frac{2l}{c}\left(1 - \frac{v^2}{c^2}\right)^{-1} \tag{2.8}$$

which, as before, can be expanded binomially to yield

$$t_2 = \frac{2l}{c}\left(1 + \frac{v^2}{c^2}\right) \tag{2.9}$$

The difference between the times taken by the transmitted and the reflected lights is therefore

$$t_2 - t_1 = \frac{lv^2}{c^3} \tag{2.10}$$

Hence the path difference between the two rays is

$$s = c(t_2 - t_1) = \frac{lv^2}{c^2} \tag{2.11}$$

If the entire apparatus is rotated through 90° about a vertical axis, the two beams interchange their roles so that the beam previously taking a longer path will now take a shorter path and vice versa. The path difference will then be in the opposite direction and so the total path difference becomes

$$\frac{lv^2}{c^2} - \left(-\frac{lv^2}{c^2}\right) = \frac{2lv^2}{c^2}$$

If this difference in the optical path corresponds to n waves of wavelength λ, we have

$$\frac{2lv^2}{c^2} = n\lambda$$

or
$$n = \frac{2lv^2}{c^2\lambda} \qquad (2.12)$$

This gives the number of fringes which should shift in the field of view due to the motion of the earth through ether.

Michelson and Morley used sodium light of wavelength $\lambda = 5.9 \times 10^{-7}$ m and an interferometer of arm length $l = 11$ m. Assuming v/c to be 10^{-4}, Eq. (2.12) gives

$$n = \frac{2 \times 11 \times 10^{-8}}{5.9 \times 10^{-7}} = 0.37$$

This predicts a shift of 0.37 of a fringe width. Such a shift, if existed, could have been easily detected with the precision of Michelson interferometer which was one-hundredth of a fringe width. But the actual shift observed by Michelson and Morley was less than even 1/20 of the expected value. Thus their experiment gave a null result, i.e., no relative motion of earth with respect to ether could be detected. Michelson and Morley thought that since the earth moves in different directions in different times of the year, it was just possible that at the time of their conducting the experiment, probably the earth had no component of velocity parallel to its surface. So they repeated their experiment at different times of the year but this also led them to the same negative result.

The above experiment was repeated by Noble and Trouton in 1904 by using electromagnetic waves (generated by electrical means) instead of visible light. · But they also met with failure. In 1921 and thereafter in 1924 Prof. Miller conducted his experiment in the Mount Wilson Laboratory. The fringe shift obtained by him was less than 1/40 of the calculated amount. So the same negative results were again obtained. Therefore, the motion of earth relative to ether could not be detected by either electromagnetic or optical experiments. This null result stunned the world of physics which had up to that time been very successful in explaining the propagation of light waves by the ether hypothesis.

Explanation of the Negative Results

There were attempts to explain the negative results obtained by Michelson-Morley and others. Two main theories proposed were

(i) ether drag theory and (ii) Lorentz-Fitzgerald contraction theory. But as we shall see none of them fully satisfactorily explained the failure of the Michelson-Morley experiment. It was only Einstein who could give the correct explanation for the negative results.

(i) ETHER DRAG THEORY

It was thought that ether was dragged along with the earth. So there was no relative motion between the earth and the ether. This amounted to saying that no fringe shift would be observed in the Michelson-Morley experiment. But then this drag, if assumed to be complete, is unable to explain the experimentally observed phenomenon of the stellar aberration. On the other hand, if this drag is assumed to be partial, as suggested by Fresnel and Fizeau, then it is difficult to explain at the same time the phenomenon of the stellar aberration and also account for the negative results of the Michelson-Morley experiment. Moreover, the partial ether drag theory proposed by Fresnel and Fizeau has been explained by Lorentz on the basis of electromagnetic theory which assumes the ether to be stationary.

(ii) LORENTZ-FITZGERALD CONTRACTION THEORY

According to this theory, moving objects appear to be contracted in the direction of their motion relative to the ether stream while the dimensions of the object perpendicular to the motion remain unaffected. If l be the original length of the object then when it moves with velocity v its length becomes

$$l' = l \sqrt{1 - v^2/c^2}$$

Following this theory, the length of the interferometer arm in the direction of the ether stream shortens to $l\sqrt{1 - v^2/c^2}$ while the length of the perpendicular arm remains unaltered.

The time t_2 in Eq. (2.8) now becomes

$$t_2 = \frac{2l}{c} \frac{\sqrt{1 - v^2/c^2}}{(1 - v^2/c^2)} = \frac{2l}{c\sqrt{1 - v^2/c^2}}$$

or

$$t_2 = \frac{2l}{c}\left(1 + \tfrac{1}{2}\frac{v^2}{c^2}\right) \qquad (2.13)$$

Comparing this with Eq. (2.4) we see that $t_1 = t_2$. Since the two times are equal, no fringe shift would be observed. However, the shortening of the interferometer arm as predicted by Lorentz-

Fitzgerald contraction theory can never be measured in practice because any measuring stick used to measure it would also be moving relative to the ether and would shorten too. Successful though was the Lorentz-Fitzgerald contraction theory in explaining the null results of Michelson-Morley experiment, it however did not have a firm theoretical basis. It was based on a somewhat *ad hoc* hypothesis.

The correct explanation of the null results ultimately came from Einstein who pointed out that it was fundamentally wrong to calculate the times t_1 and t_2 using the Galilean transformation equation for velocity. It occurred to Einstein that the velocity of earth or any observer through space does not make any difference in the speed of light which is therefore constant and independent of the relative motion between the source and the observer.

The search for a fixed reference frame with respect to which Newton's laws would hold exactly, was the aim of the Michelson-Morley experiment. Since this experiment failed it was concluded that motion of earth through ether was undetectable and that there is no point in speaking of a frame which is absolutely at rest. In fact there could be no absolute velocities for every velocity would have to be measured relative to an origin that might and probably would be moving. As there can be no preferred reference frame, all frames are equivalent for the description of the physical laws of nature.

2.4. THE POSTULATES OF THE SPECIAL RELATIVITY

Einstein made the constancy of the velocity of light a cornerstone of his special or restricted theory of relativity which he proposed in 1905. This theory deals with inertial frames, i.e., frames of reference which have a constant linear velocity relative to one another. The general theory of relativity given by Einstein in 1915, deals with non-inertial or accelerated frames. We shall deal here only with his special theory of relativity. Einstein based his theory on two fundamental postulates which can be stated as follows:

(*i*) *The laws of physics remain the same when stated in terms of frames of reference moving with uniform rectilinear motion with respect to one another.*

(*ii*) *Velocity of light in vacuum is constant and is independent not only of the direction of propagation of light but also of the relative motion between the source and the observer.*

2.5. LORENTZ–EINSTEIN TRANSFORMATIONS

Galilean transformation equations were derived in the preceding chapter assuming space and time to be absolute. But according to Einstein there is no concept like "absolute". All motions in the universe are relative. A modified set of relativistic transformation equations was derived by H.A. Lorentz (though in a slightly less general form) from a study of Maxwell's equations and independently by Einstein assuming the invariance of the speed of light (second postulate of the special theory of relativity). These relativistic transformation equations are called Lorentz-Einstein transformations or simply the Lorentz transformations.

Consider two reference frames S and S' with S' moving with a linear velocity v along the $+x$ direction relative to frame S (Fig. 2.3). The observers making measurements are located at the origins O and O' of the respective frames. Let the origins of the two frames coincide initially at $t = t' = 0$. Starting at this time let a spherical

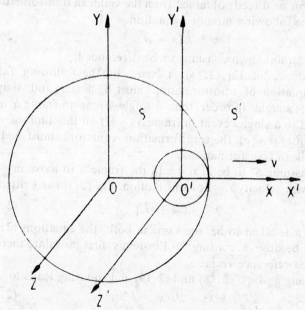

Fig. 2.3. Reference frames S and S' in relative motion.

light wave spread out from the origin. In course of time the wave will spread out in the form of a sphere.

The equation of the expanding sphere as seen by the observer O is

$$x^2 + y^2 + z^2 = c^2 t^2 \qquad (2.14)$$

Similarly, observer O' writes his equation for the sphere in the form

$$x'^2 + y'^2 + z'^2 = c^2 t'^2 \qquad (2.15)$$

Since the motion is along the x direction the coordinates y, y', z and z' remain unaffected and so

$$y = y', \; z = z' \qquad (2.16)$$

Subtracting Eq. (2.15) from Eq. (2.14) and using Eq. (2.16) we obtain

$$x^2 - x'^2 = c^2(t^2 - t'^2) \qquad (2.17)$$

According to the Galilean transformations assuming space and time to be absolute, x' is related to x by the equation $x' = x - vt$. But according to the Einsteinian relativity space and time measurements depend on the observer. Experiments and theory reveal that transformation equations assuming space and time to be non-absolute, can be directly obtained from the Galilean transformations by using the following modified equation

$$x' = k(x - vt) \qquad (2.18)$$

where k is an unknown constant to be determined.

Our choice of Eq. (2.18) is based on the following facts: (1) The equation of transformation must be linear and simple. Linearity is sought in order that a single event in frame S may correspond to a single event in frame S'. (2) In the limit of very small velocities ($v \ll c$), the transformation equations must reduce to the Galilean transformations.

Now assume S' to be at rest and the frame S to move in $-x'$ direction with velocity v. For this motion, Eq. (2.18) is written as

$$x = k(x' + vt') \qquad (2.19)$$

The factor k is taken to be the same in both the equations (2.18) and (2.19) because according to Einstein's first postulate there is no preferred reference frame.

Rearranging Eqs. (2.18) and (2.19) and squaring leads to

$$x'^2 + k^2 x^2 - 2kxx' = k^2 v^2 t^2 \qquad (2.20)$$

$$x^2 + k^2 x'^2 - 2kxx' = k^2 v^2 t'^2 \qquad (2.21)$$

Subtracting Eq. (2.21) from Eq. (2.20) we have

$$(x'^2 - x^2) + k^2(x^2 - x'^2) = k^2v^2(t^2 - t'^2)$$

or

$$(x^2 - x'^2)(k^2 - 1) = k^2v^2(t^2 - t'^2) \qquad (2.22)$$

Comparing this with Eq. (2.17) we obtain

$$\frac{k^2v^2}{k^2 - 1} = c^2$$

which can be simplified further to yield

$$k = \frac{1}{\sqrt{1 - v^2/c^2}} \qquad (2.23)$$

Eliminating x' between Eqs. (2.19) and (2.18) and solving for t' gives

$$t' = k(t - vx/c^2) \qquad (2.24)$$

Eqs. (2.18), (2.16) and (2.24) define the complete set of Lorentz transformation equations:

$$x' = k(x - vt)$$
$$y' = y$$
$$z' = z$$
$$t' = k(t - vx/c^2) \qquad (2.25)$$

The inverse Lorentz transformations are readily obtained by replacing v by $-v$ in the above equations and interchanging primed and unprimed quantities. We have

$$x = k(x' + vt')$$
$$y = y'$$
$$z = z'$$
$$t = k(t' + vx'/c^2) \qquad (2.26)$$

It can be shown that Lorentz transformations reduce to the classical Galilean transformations in the limit of very small velocities ($v \ll c$). In this limit, $k \simeq 1$ and $v/c^2 \to 0$. Hence the transformation equations (2.25) reduce to

$$x' = x - vt$$
$$y' = y$$
$$z' = z$$
$$t' = t$$

which are the familiar Galilean transformations. The classical physics can, therefore, be regarded as a special case of the relativity physics. An obvious conclusion is that the relativistic effects become significant only at very high speeds approaching the speed of light.

2.6. RELATIVITY OF SPACE AND TIME AND THE SIMULTANEITY OF EVENTS

We shall now discuss some consequences of Lorentz transformation equations. We shall show that moving rods contract in their direction of motion while moving clocks appear to run slow. Also, two events which are simultaneous with respect to one observer may no longer be simultaneous with respect to another observer who is in relative motion.

(i) LORENTZ CONTRACTION

Consider a rod placed along the x axis of the reference frame S (Fig. 2.4). Let the x coordinates of the ends of the rod as measured by the observer at O be x_1 and x_2. Then we have for the length of the rod l (also called *proper* length) in the frame S

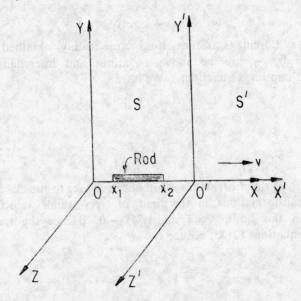

Fig. 2.4.   Rod placed in system S.

$$x_2 - x_1 = l \tag{2.27}$$

Let the observer at O' measure the x coordinates of the ends of the rod at the same instant of time t' and let these coordinates be x_1' and x_2'. We have

$$x_2' - x_1' = l' \tag{2.28}$$

For the coordinates of the two ends of the rod we have using the inverse Lorentz transformation (Eq. 2.19):

$$x_1 = k(x_1' + vt')$$
$$x_2 = k(x_2' + vt') \tag{2.29}$$

from which we obtain

$$x_2 - x_1 = k(x_2' - x_1') \tag{2.30}$$

But $x_2' - x_1' = l'$ and $x_2 - x_1 = l$, the proper length of the rod. We thus have

$$l' = l/k$$

or
$$l' = l\sqrt{1 - v^2/c^2} \tag{2.31}$$

Eq. (2.31) tells that the length of the rod as seen by observer at O' is less than the corresponding length as measured by the observer at O. We may say that the length of the rod appears to be contracted in the direction of motion. Note that the *proper* length is the *maximum* length that can be measured by an observer.

(ii) TIME DILATION

Consider a clock placed in position x in the frame of reference S (Fig. 2.5). The clock emits signals at successive instants of time t_1 and t_2. Let these time instants, as measured by the observer at O', be t_1' and t_2'. Then we have from the Lorentz transformation for time (Eq. 2.24),

$$t_1' = k(t_1 - vx/c^2)$$
$$t_2' = k(t_2 - vx/c^2) \tag{2.32}$$

Subtraction yields

$$t_2' - t_1' = k(t_2 - t_1)$$

or
$$\Delta t' = k\,\Delta t \tag{2.33}$$

Thus the time interval to the observer at O' appears to be elongated or dilated. From this we can conclude that moving clocks appear to run slow. Note that the time interval measured by an observer stationary with respect to the clock (in this case it is Δt) is the

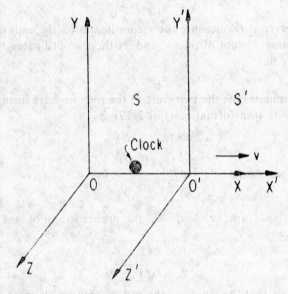

Fig. 2.5. Clock placed in system S.

proper time interval. This is the *minimum* time interval which an observer can measure between two events.

(iii) PRINCIPLE OF SIMULTANEITY

Consider two events P_1 and P_2 to occur in frame S (Fig. 2.6). These events are specified by the coordinates (x_1, y_1, z_1, t_1) and (x_2, y_2, z_2, t_2) respectively. Let the corresponding coordinates of the two events as seen by observer at O' be (x_1', y_1', z_1', t_1') and (x_2', y_2', z_2', t_2') respectively. Then we have

$$t_1' = k\left(t_1 - \frac{vx_1}{c^2} \right) \qquad (2.34)$$

$$t_2' = k\left(t_2 - \frac{vx_2}{c^2} \right) \qquad (2.35)$$

which on subtraction gives

$$t_1' - t_2' = k\,(t_1 - t_2) + \frac{kv}{c^2}\,(x_2 - x_1) \qquad (2.36)$$

If the events P_1 and P_2 are simultaneous with respect to frame S, they must occur at the same time, i.e., $t_1 = t_2$. So we have from Eq. (2.36)

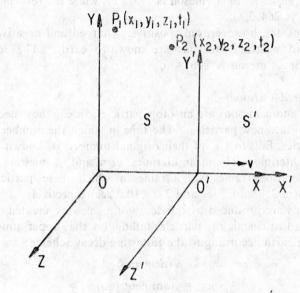

Fig. 2.6. Events P_1 and P_2 in frame S.

$$t_1' - t_2' = \frac{kv}{c^2}(x_2 - x_1) \neq 0$$

or $\qquad\qquad t_1' \neq t_2' \qquad\qquad\qquad\qquad$ (2.37)

Thus we see that the two events which are simultaneous with respect to S are no longer simultaneous with respect to S'. So, the principle of simultaneity is not an absolute concept, i. e., the simultaneity of two events depends on the observer or the frame of reference.

Verification of Time Dilation

An experimental support has been lent to the time dilation theory by considering the decay characteristics of certain subatomic particles called mesons. Mesons ("meso" is a greek term which means middle) have masses intermediate between the electron mass and the proton mass. Three types of mesons are known to exist, namely, π, μ and K mesons. π mesons (also called primary mesons) are either positively charged, negatively charged or neutral (π^+, π^-, π^0). The charges carried by π^+ and π^- mesons are the same in magnitude as that carried by an electron. The rest mass of either π^+ or π^- meson is 273.1 m_e while the rest mass of π^0 meson is 264.3 m_e.

μ mesons are, however, only positively charged and negatively charged. No neutral μ mesons are known to exist. The rest mass of μ^+ or μ^- meson is 206.8 m_e.

Decay of π and μ Mesons

Both π and μ mesons are unstable particles, i. e., they decay or transform into new particles. The time in which the number of these particles falls to $1/e$ of their original number is known as their mean lifetime. The mean lifetimes of π and μ mesons in their rest or proper frames (i.e., a frame in which these particles are at rest) are 2.6×10^{-8} sec and 2.2×10^{-6} sec respectively.

μ mesons are produced by the decay of π mesons created by the action of fast cosmic ray particles falling on the upper atmosphere of the earth according to the following decay scheme:

$$\pi^+ \rightarrow \mu^+ + \nu \text{ (neutrino)}$$

$$\pi^- \rightarrow \mu^- + \bar{\nu} \text{ (antineutrino)}$$

The μ mesons further decay into electron (or positron) and neutrino-antineutrino pairs as follows :

$$\mu^+ \rightarrow e^+ + \nu + \bar{\nu}$$

$$\mu^- \rightarrow e^- + \nu + \bar{\nu}$$

μ mesons created high in the atmosphere have a typical speed of 2.994×10^8 m/sec or $0.998c$. With this velocity μ mesons, before decaying, are expected to travel a distance of roughly 659 m. The earth is at a distance of about 10 km from that part of the atmosphere where μ mesons are created. Thus apparently there seems to be no possibility for the μ mesons to reach the earth. This is, however, contrary to actual observations as a significant number of μ mesons have been detected on earth.

The above paradox can be resolved only by taking into account the dilation in the meson's lifetime as seen by an observer on earth. As seen by the earth-bound observer, the life time of meson will be extended to the value

$$t = t_0/\sqrt{1 - v^2/c^2} = 2.2 \times 10^{-6}/\sqrt{1 - (.998)^2} = 35 \times 10^{-6} \text{ sec}$$

In this time the distance actually travelled by μ meson will be about 10.5 km. Thus time dilation is able to explain the meson paradox. This establishes the validity of the time dilation theory.

Time dilation theory has also been tested in the laboratory by measuring the lifetimes of π mesons created by bombarding the target of an accelerator by high energy protons. The results obtained once again prove the validity of the dilation theory.

QUESTIONS AND PROBLEMS

2.1. What was the motivation behind Michelson-Morley experiment ? Describe the experiment and show how the negative results obtained therefrom are explained. What conclusions do you draw from the experiment regarding the existence of ether ?

2.2. Starting from the basic postulates of the special theory of relativity deduce the Lorentz transformation equations. Show that for very small velocities these equations reduce to the classical Galilean transformations.

2.3. What is proper length of an object? Discuss using Lorentz transformations the contraction suffered by a moving object in the direction of its motion. How is Lorentz-Fitzgerald contraction able to account for the null result of Michelson-Morley experiment?

2.4. What do you understand by time dilation at relativistic speeds? What is proper time interval? Give some evidence in support of the time dilation theory.

2.5. What does the simultaneity of two events mean? Discuss the relativity of simultaneity and show that two events which are simultaneous with respect to a reference frame are not in general simultaneous with respect to another frame which has a constant linear velocity relative to the first frame.

2.6. A metre stick is projected into space at so great a speed that its length appears contracted to only 50 cm. How far is it going in m/sec?   (2.598×10^8 m/sec)

2.7. A rocket ship is 100 m long on the ground. When it is in flight, its length is 99 m to an observer on the ground. What is its speed?   (4.2×10^7 m/sec)

2.8. A rod is moving with a velocity 0.6 times the velocity of light with respect to the laboratory. If an observer situated in the laboratory measures its length as 1 m, calculate the proper length of the rod.

(1.25 m)

2.9. Calculate the percentage contraction of a rod moving with a velocity $0.8c$ in a direction inclined at $60°$ to its own length.

(8.4%)

[Hint: The length of the rod in its proper frame along and perpendicular to the direction of motion will be $l_0 \cos 60°$ and $l_0 \sin 60°$ respectively, l_0 being the proper length of the rod. If x and y axes be taken respectively along and perpendicular to the direction of motion, then as seen by the moving observer, $l_x = l_0 \cos 60° \sqrt{1 - v^2/c^2}$ and $l_y = l_0 \sin 60°$]

2.10. A clock gives correct time in its rest frame. With what speed must the clock move relative to an observer so that it may seem to lose 1 minute in 24 hours?

(1.1×10^7 m/sec)

2.11. In the laboratory the mean lifetime of a particle moving with speed 2.8×10^8 m/sec is found to be 2.5×10^{-7} sec. Calculate the mean proper lifetime of the particle.

(0.9×10^{-7} sec)

2.12. π mesons moving with velocity $0.99c$ decay with mean lifetime of 2.6×10^{-8} sec as measured in their proper or rest frame. What is their lifetime as measured by the laboratory observer?

(1.84×10^{-7} sec)

2.13. A certain particle has a lifetime of 10^{-7} sec when measured at rest. How far does it go before decaying if its speed is $0.99c$ when it is created?

(210.4 m)

2.14. A particle with a mean proper lifetime of 1 μsec moves through the laboratory at 2.7×10^8 m/sec. (a) What is its lifetime as measured by observers in the laboratory? (b) If it was manufactured in the target of an accelerator, how far does it go, on the average, in the laboratory before disintegrating?

((a) 2.3 μsec, (b) 6.2×10^2 m)

3. Relativistic Dynamics

3.1. INTRODUCTION

Certain relativistic effects were dealt with in the preceding chapter. The present chapter will examine the dynamics of a relativistically moving particle. The Galilean transformation for velocity ($u' = u - v$) is not valid for relativistic velocities. We shall derive the relativistic velocity transformation formulae and shall show them to reduce to the classical velocity transformations in the limit of low velocities ($v \ll c$).

In Newtonian mechanics the basic physical quantity is the mass or inertia. However, in relativistic mechanics the mass of a particle varies with velocity although its momentum is taken to be conserved. As will be seen, the momentum in relativistic dynamics can be represented by a four-vector. A four-vector, contrary to three-dimensional vector in the ordinary space, has four components. A concept of considerable importance is the equivalence of mass and energy, namely, that mass and energy are not independent entities but are inter-convertible. We shall establish the mass-energy equivalence relation due to Einstein. This principle is the basis of energy released in the fusion process taking place in sun or the fission process in an atom bomb, due to the conversion of matter or mass into energy.

3.2. RELATIVISTIC ADDITION OF VELOCITIES

Let S and S' be two inertial reference frames with S' moving with respect to S with a constant linear velocity v along the $+ x$ direction. Consider a particle moving relative to both frames so that it has velocity components u'_x, u'_y, u'_z with respect to frame S'

and components u_x, u_y, u_z with respect to S. Using the definition of velocity, these components can be expressed as

$$u_x' = \frac{dx'}{dt'}; \quad u_x = \frac{dx}{dt}$$

$$u_y' = \frac{dy'}{dt'}; \quad u_y = \frac{dy}{dt} \qquad (3.1)$$

$$u_z' = \frac{dz'}{dt'}; \quad u_z = \frac{dz}{dt}.$$

We now differentiate the Lorentz transformation equations (Eq. 2.25) to obtain

$$dx' = \frac{dx - vdt}{\sqrt{1 - v^2/c^2}}$$

$$dy' = dy$$

$$dz' = dz \qquad (3.2)$$

$$dt' = \frac{dt - (v/c^2)\,dx}{\sqrt{1 - v^2/c^2}}$$

If we divide each of the differentials dx', dy', dz' by dt', we can at once get

$$\frac{dx'}{dt'} = \frac{(dx/dt) - v}{1 - \frac{v}{c^2}(dx/dt)}$$

$$\frac{dy'}{dt'} = \frac{(dy/dt)\sqrt{1 - v^2/c^2}}{1 - \frac{v}{c^2}(dx/dt)} \qquad (3.3)$$

$$\frac{dz'}{dt'} = \frac{(dz/dt)\sqrt{1 - v^2/c^2}}{1 - \frac{v}{c^2}(dx/dt)}$$

Using Eq. (3.1) we have ultimately

$$u_x' = \frac{u_x - v}{1 - vu_x/c^2}$$

$$u_y' = \frac{u_y\sqrt{1 - v^2/c^2}}{1 - vu_x/c^2} \qquad (3.4)$$

$$u_z' = \frac{u_z\sqrt{1 - v^2/c^2}}{1 - vu_x/c^2}$$

The inverse velocity transformations expressing u_x, u_y, u_z in terms of u'_x, u'_y, u'_z can be obtained conveniently by changing the sign of v and interchanging primed and unprimed quantities. We have

$$u_x = \frac{u'_x + v}{1 + vu'_x/c^2}$$

$$u_y = \frac{u'_y \sqrt{1 - v^2/c^2}}{1 + vu'_x/c^2}$$

$$u_z = \frac{u'_z \sqrt{1 - v^2/c^2}}{1 + vu'_x/c^2}$$

(3.5)

The non-relativistic forms of Eqs. (3.4) and (3.5) are readily derived by setting the limit $v/c \to 0$ and these equations are seen to reduce to the Galilean results.

We can derive certain conclusions concerning the velocity of light from Eqs. (3.4) and (3.5). If the moving object is a photon moving in the S' frame with velocity c in the $+x$ direction, we have

$$u_x = \frac{c + v}{1 + \frac{cv}{c^2}} = c$$

(3.6)

We see that the velocity of the photon in the second frame is also c. Thus, we conclude that the velocity of light in vacuum is an absolute quantity independent of the frame of reference chosen.

If the frame S' translates with velocity c with respect to the frame S we set $v = c$ in Eq. (3.6) and obtain

$$u_x = \frac{c + c}{1 + \frac{c^2}{c^2}} = c$$

This shows that addition of any velocity to the velocity of light merely reproduces the velocity of light. Thus velocity of light is the maximum velocity that can be attained in nature by any material body.

It may be of general interest to show that the relativistic addition of two velocities each less than c can never result in a velocity greater than c. To prove this we rewrite first of Eqs. (3.5) in the form

$$u_x = \left[\frac{(u'_x/c) + (v/c)}{1 + (v/c)(u'_x/c)}\right] c$$

$$= \left[1 - \frac{(1 - u'_x/c)(1 - v/c)}{1 + (v/c)(u'_x/c)}\right] c$$

(3.7)

As both u'_x and v are less than c it is clear from Eq. (3.7) that $u_x < c$ which proves our result.

3.3. RELATIVITY OF MASS

While discussing the effect of relativistic motion on length and time, we found in the preceding chapter that length and time, contrary to classical Newtonian mechanics, can no longer be regarded as absolute. We shall now show that, according to the relativity theory, mass too is not an absolute quantity, i.e., its value depends on the frame of reference or the observer making measurements. In fact, as will be shown, the mass of a moving object increases in motion. We shall derive the mass variation formula assuming that the law of conservation of momentum holds even for relativistic speeds.

We consider two frames of reference S and S', the latter moving with uniform velocity v along the $+x$ direction with respect to frame S. Imagine two particles each having mass m'

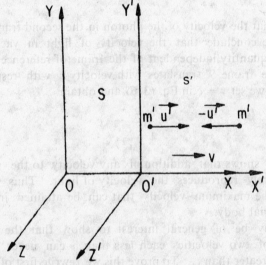

Fig. 3.1. Collision of particles in frame S' moving towards each other.

moving with velocities $\vec{u}'$ and $-\vec{u}'$ along the $+x$ direction of the frame S'. Let these particles collide head-on and after collision coalesce into a single particle. So we have by applying the law of conservation of momentum in the primed frame S'

$$m'u' - m'u' = 2m'v_{ai} \tag{3.8}$$

where v_{ai} is the velocity of the coalesced mass after impact.

From Eq. (3.8) we have

$$v_{ai} = 0$$

This means that, after collision, the coalesced mass is at rest in the S' frame. However, an observer at O in the S frame will see the coalesced mass as moving with velocity v along the $+ x$ direction.

If m_1 and m_2 are the respective masses of the particles, as seen by the observer in the frame S, and u_1 and u_2 their velocities before impact, then by applying the momentum conservation principle in the unprimed frame S

$$m_1 u_1 + m_2 u_2 = (m_1 + m_2) v$$

Rearranging, we get

$$m_1 (u_1 - v) = m_2 (v - u_2)$$

or

$$\frac{m_1}{m_2} = \frac{v - u_2}{u_1 - v} \tag{3.9}$$

Using the formulae for the relativistic composition of velocities, Eq. (3.5), we can write

$$u_1 = \frac{u' + v}{1 + u'v/c^2}, \quad u_2 = \frac{- u' + v}{1 - u'v/c^2} \tag{3.10}$$

From Eq. (3.10), we can obtain the values for $u_1 - v$ and $v - u_2$ in the form

$$u_1 - v = \frac{u' (1 - v^2/c^2)}{1 + u'v/c^2}, \quad v - u_2 = \frac{u' (1 - v^2/c^2)}{1 - u'v/c^2} \tag{3.11}$$

Substituting in Eq. (3.9)

$$\frac{m_1}{m_2} = \frac{1 + u'v/c^2}{1 - u'v/c^2} \tag{3.12}$$

It will be convenient if we evaluate the factors $1 - u_1^2/c^2$ and $1 - u_2^2/c^2$ beforehand. These factors readily simplify to

$$1 - \frac{u_1^2}{c^2} = \frac{(1 - u'^2/c^2)(1 - v^2/c^2)}{(1 + u'v/c^2)^2} \tag{3.13}$$

and

$$1 - \frac{u_2^2}{c^2} = \frac{(1 - u'^2/c^2)(1 - v^2/c^2)}{(1 - u'v/c^2)^2} \tag{3.14}$$

Dividing Eq. (3.13) by Eq. (3.14) we obtain

$$\frac{1 - \dfrac{u_2{}^2}{c^2}}{1 - \dfrac{u_1{}^2}{c^2}} = \frac{(1 + u'v/c^2)^2}{(1 - u'v/c^2)^2}$$

or
$$\frac{\sqrt{1 - u_2{}^2/c^2}}{\sqrt{1 - u_1{}^2/c^2}} = \frac{1 + u'v/c^2}{1 - u'v/c^2} \qquad (3.15)$$

Combining Eqs. (3.12) and (3.15)

$$\frac{m_1}{m_2} = \frac{\sqrt{1 - u_2{}^2/c^2}}{\sqrt{1 - u_1{}^2/c^2}}$$

or
$$m_1 \sqrt{1 - u_1{}^2/c^2} = m_2 \sqrt{1 - u_2{}^2/c^2}$$

Since each side of this equation is independent, this can be only true if each side is a constant. So we set

$$m_1 \sqrt{1 - u_1{}^2/c^2} = m_2 \sqrt{1 - u_2{}^2/c^2} = m_0 \text{ (constant)} (3.16)$$

The form of this equation suggests that if there is a mass m moving with velocity v it is given by

$$m \sqrt{1 - v^2/c^2} = m_0$$

or
$$m = m_0/\sqrt{1 - v^2/c^2} \qquad (3.17)$$

where m_0 is called the rest mass of the particle, i.e., mass of the particle when it is at rest.

Equation (3.17) is the expression for the relativistic change of mass. It tells that a body of rest mass m_0 has its mass increased by the factor $1/\sqrt{1 - v^2/c^2}$ when it moves with velocity v relative to an observer. From Eq. (3.17) it can also be seen that as $v \to c$, $m \to \infty$ which means that an infinite force will be needed to accelerate the body to the velocity of light since the mass then becomes infinity. But there are neither infinite forces nor infinite masses in the universe. This once again lays emphasis on the fact that no material object can attain a speed equal to or greater than the speed of light.

It should be pointed out that objects in everyday life have velocities very small compared to the velocity of light. Therefore, the predicted effects of length contraction, time dilation, and mass variation cannot be observed at ordinary speeds. Only atomic particles such as electron, proton, deuteron etc. have sufficiently high speeds for relativistic effects to be measurable.

3.4. VERIFICATION OF MASS VARIATION FORMULA: BUCHERER EXPERIMENT

The first experimental test of the mass transformation formula was made by Bucherer in 1908 using electrons from a radioactive source.

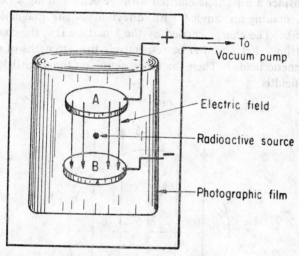

Fig. 3.2. Diagram showing Bucherer's apparatus. The magnetic lines of force are parallel to the discs.

The apparatus consists of two disc shaped plates A and B (Fig. 3.2). A small grain of radium fluoride, which acts as a radioactive source emitting β rays (or electrons), is placed between the plates. An electric field is applied between the plates A and B by connecting them to the positive and negative terminals of a source of high potential difference. A magnetic field is also applied by means of the pole pieces of an electromagnet. The magnetic lines of force are taken to lie in the plane of the discs. A cylindrical photographic film coaxial with the discs is also fitted with the apparatus. The whole apparatus is enclosed in a vacuum chamber.

The β particles emitted from the source are subjected to two forces, one due to the electric field and another due to the magnetic field. The particles are actually emitted along all the radii of the discs making different angles with the magetic lines of force depending on their velocities. The separation between the plates is small (about 0.25 mm). Therefore, in order that β particles might escape from between the discs and strike the photographic film, the forces on them due to the electric and the magnetic fields must be equal and opposite. We now calculate these forces.

Consider a β particle emitted with velocity v along a radius of the disc, making an angle θ with direction of the magnetic field (Fig. 3.3). The charge carried by the β particle is e, the charge on the electron. Let E and H be respectively the strengths of electric and magnetic fields. Then the force acting on the β particle in the electric field is

$$F_E = eE \qquad (3.18)$$

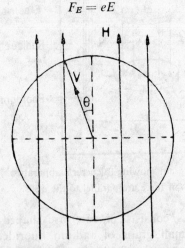

Fig. 3.3. Configuration showing the upper surface of the disc and the magnetic lines of force.

The force experienced by the β particle in the magnetic field is given by

$$F_H = e\,(\vec{v} \times \vec{H}) = evH \sin\theta \qquad (3.19)$$

Both the forces F_E and F_H act perpendicular to the plane of the discs but are in opposite directions. Thus while F_E acts in the upward direction, the direction of F_H is downward (Fig. 3.4). So,

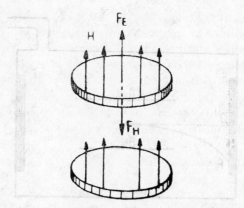

Fig. 3.4. The directions of electric and magnetic forces.

if $F_E > F_H$, β particle will be attracted towards plate A; if $F_H > F_E$ it will be attracted towards B. In either of these two cases β particles obviously cannot emerge from between the plates. However, if $F_E = F_H$, no net force will act on the β particle which will as a consequence escape from between the discs and strike the photographic film. This will happen when

$$eE = evH \sin \theta$$

or
$$v = E/H \sin \theta \qquad (3.20)$$

According to Eq. (3.20) β particles emitted at an angle θ with respect to the magnetic field and moving with a velocity v will be able to emerge from between the plates. Any other particle emitted at angle θ but moving with velocity different from v will not be able to escape and consequently will be trapped between the discs.

The β particles possessing velocities satisfying Eq. (3.20) will have no net force acting on them and so they will start moving toward the photographic film along the dotted line as shown in Fig. 3.5. But as soon as β particles cross the region of the discs they get subjected to a magnetic field due to which they start moving in a circular path. Thus β particles will not strike the photographic film at O (which they should have done had no force acted on them) but at P. The deflection OP caused by the magnetic field, called magnetic deflection, can be calculated as follows:

The acceleration experienced by β particles in the magnetic field is $\alpha = F_H/m = evH \sin \theta/m$. If t stands for the time which β

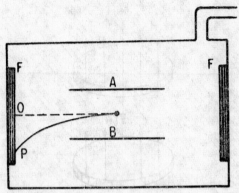

Fig. 3.5.   Deflection of β particle in the magnetic field.
FF represents the photographic film.

particles would take in moving from the edge of the discs to the film, we have

$$OP = \tfrac{1}{2}\alpha\, t^2$$

$$= \tfrac{1}{2}\left(\frac{evH\sin\theta}{m}\right)t^2 \tag{3.21}$$

Note that the initial velocity component of β particles along the direction OP was zero as they had a velocity component only along the dotted path.

The time t in (3.21) is given by

$$t = a/v \tag{3.22}$$

where a is the distance from the edge of the discs to the film.

Combining Eqs. (3.21) and (3.22) we obtain

$$OP = \tfrac{1}{2}\,(evH\sin\theta/m)\,(a^2/v^2)$$

which gives for the ratio e/m the value

$$\frac{e}{m} = \frac{2\,OPv}{a^2\,H\sin\theta} \tag{3.23}$$

Thus e/m for β particles of different velocities can be calculated using Eq. (3.23). Particles with different velocities will strike the film at different points. When the film is finally stretched out, a curve of the form shown in Fig. 3.6 is obtained. Note that the other half of the trace is obtained by reversing the directions of electric and magnetic fields. From the curve, magnetic deflections such as OP and OP', can be known. Thus e/m can be calculated for different velocities.

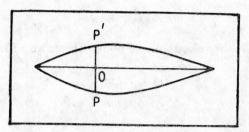

Fig. 3.6. Trace on the photographic film.

Bucherer made $E/H = c/2$ in his experiment, c being the velocity of light. Taken with Eq. (3.20) this condition gives $v/c = 1/2 \sin \theta$. If $v = c$, then $\sin \theta = \frac{1}{2}$ so that we have $\theta = 30°$ and 150°. Therefore, only those β particles which are emitted initially at an angle lying between 30° and 150° with respect to the magnetic field will be able to emerge from between the plates and strike the photographic film.

From the experiment e/m for β particles of different velocities were determined. It was found that the ratio $e/m\sqrt{1 - v^2/c^2}$ remained constant signifying that e/m_0 was the same, as it should, for all β particles irrespective of their velocities. Thus the validity of the mass variation formula was established by means of Bucherer's experiment.

3.5. EINSTEIN'S MASS-ENERGY RELATION

Newton's second law expressing force as product of mass and acceleration is no longer valid in relativistic dynamics because mass is no longer a constant quantity. Therefore, Newton's law must be written in a more general form as

$$F = \frac{d}{dt}(mv) \qquad (3.24)$$

The kinetic energy of a body moving with velocity v is defined as the work done by a force in accelerating the body from rest to the velocity v. If a force F acting on a body produces a displacement dx in it, then the change in the kinetic energy of the body will be the work done which is force times distance, i.e.,

$$dE_k = Fdx \qquad (3.25)$$

Combining Eqs. (3.24) and (3.25) we obtain

$$dE_k = \frac{d}{dt}(mv)\, dx \tag{3.26}$$

But $dx/dt = v$, the velocity of the body. Hence we write Eq. (3.26) in the form

$$dE_k = v\, d(mv)$$
$$= v(v\, dm + m\, dv)$$
$$= v^2\, dm + mv\, dv \tag{3.27}$$

This expression can be simplified using the mass variation formula $m = m_0/\sqrt{1 - v^2/c^2}$, which after squaring and rearranging becomes

$$m^2c^2 = m^2v^2 + m_0^2c^2 \tag{3.28}$$

Differentiating both sides of Eq. (3.28) recalling that m_0 and c are constants leads to

$$2mc^2\, dm = 2mv^2\, dm + 2vm^2\, dv$$

or $\qquad\qquad c^2\, dm = v^2\, dm + mv\, dv \tag{3.29}$

Comparing Eqs. (3.27) and (3.29)

$$dE_k = c^2\, dm \tag{3.30}$$

Let the body initially at rest acquire a velocity v under the action of the force. So the mass changes from its rest value m_0 to m, the mass in motion, and the kinetic energy increases from 0 to E_k. Integrating Eq. (3.30)

$$\int_0^{E_k} dE_k = \int_{m_0}^{m} c^2\, dm$$

or $\qquad\qquad E_k = c^2(m - m_0) \tag{3.31}$

Eq. (3.31) tells that the kinetic energy of a body is equal to the product of its mass change $(m - m_0)$ and the square of the velocity of light. Thus the increase in the kinetic energy of a body is a direct consequence of its mass change with velocity.

We can rewrite Eq. (3.31) in the form

$$mc^2 = E_k + m_0c^2$$

If we interpret mc^2 as the total energy E possessed by the body we have

$$E = E_k + m_0c^2 \tag{3.32}$$

When the body is at rest its kinetic energy E_k becomes zero. It follows from Eq. (3.32) that the body still possesses the energy $E_0 = m_0 c^2$ which may be called the rest energy of the body corresponding to its rest mass. Eq. (3.32) may then be written as

$$E = E_0 + E_k \tag{3.33}$$

The rest energy may be regarded as a form of internal store of energy in the body. Thus we see that in addition to the familiar forms of energy such as kinetic, potential, electromagnetic and thermal, there is yet another kind called mass energy.

Using Eq. (3.31), Eq. (3.33) may be written as

$$E = m_0 c^2 + c^2(m - m_0)$$

or $\qquad\qquad E = mc^2 \tag{3.34}$

which is the famous Einstein's mass-energy relation.

It can be shown that the relativistic expression for kinetic energy reduces to the classical expression $\frac{1}{2}mv^2$ for very low velocities.

Using the mass variation formula $m = m_0/\sqrt{1 - v^2/c^2}$ in Eq. (3.31) we can write the relativistic expression for kinetic energy in the form

$$E_k = \frac{m_0 c^2}{\sqrt{1 - v^2/c^2}} - m_0 c^2 \tag{3 35}$$

Expanding the first term on right of this equation with the help of binomial expansion for the non-relativistic limit ($v \ll c$) gives

$$E_k = m_0 c^2(1 + \tfrac{1}{2} v^2/c^2) - m_0 c^2$$
$$= \tfrac{1}{2}m_0 v^2 \tag{3.36}$$

Evidences in Support of Mass-Energy Relationship

A number of evidences exist which prove the validity of the mass-energy relation, $E = mc^2$. We shall give here only a few examples.

(a) ELECTRON-POSITRON ANNIHILATION AND PAIR PRODUCTION

A positron is an elementary particle which is just like the electron except that it has a positive charge. When an electron and a positron come close together they annihilate with the production of γ rays. These γ rays are found to carry energy which is equal to the combined rest energies of the electron and the positron.

The validity of mass-energy equation is also tested in the pair production process where a γ ray photon disappears creating an electron-positron pair.

(b) NUCLEAR FUSION AND FISSION PROCESSES

The confirmation of the equivalence principle is most directly established by the role it plays in fusion and fission processes. In nuclear fission a heavier nucleus breaks into two fragments, the total rest mass of both fragments being slightly less than the rest mass of the original nucleus. The difference of the two masses Δm appears in the form of energy in accordance with the relation $\Delta E = \Delta mc^2$. This principle of energy release is used in an atom bomb. In nuclear fusion, two light nuclei combine to form another nucleus whose mass is slightly less than the combined masses of the two original nuclei. The difference in masses again appears in the form of energy.

The production of energy in sun and some other stars is also a direct consequence of the validity of mass-energy relation.

3.6. RELATIVISTIC RELATION BETWEEN ENERGY AND MOMENTUM

Consider a body of rest mass m_0 moving with velocity u. The energy and the momentum of the body are given by the relations

$$E = m_0 c^2 / \sqrt{1 - u^2/c^2} \qquad (3.37)$$

$$p = m_0 u / \sqrt{1 - u^2/c^2} \qquad (3.38)$$

Eq. (3.38) may be written in the form

$$p = \frac{m_0 c^2 u}{c^2 \sqrt{1 - u^2/c^2}} = \frac{Eu}{c^2} \qquad (3.39)$$

Now squaring Eq. (3.37) and rearranging

$$E^2(1 - u^2/c^2) = m_0^2 c^4 \qquad (3.40)$$

Substituting the value of $\dfrac{u}{c}$ ($= pc/E$) from Eq. (3.39) into Eq. (3.40) we obtain

$$E^2(1 - p^2 c^2/E^2) = m_0^2 c^4$$

or $$E^2 = p^2 c^2 + m_0^2 c^4 \qquad (3.41)$$

which expresses the energy of a body in terms of its momentum.

Energy is often expressed in MeV (1 MeV $= 1.6 \times 10^{-13}$ J).

From Eq. (3.41) it follows that momentum can be expressed in terms of a new unit MeV/c. The value of this unit is

$$1 \text{ MeV}/c = \frac{1 \cdot 6 \times 10^{-13} \text{ J}}{3 \times 10^{8} \text{m/sec}} = 5.3 \times 10^{-22} \text{ kg-m/sec.}$$

3.7. MOMENTUM ENERGY FOUR-VECTOR

In the beginning of this chapter we remarked that in relativistic dynamics momentum is expressible in terms of a four-vector. We shall now prove this.

Unlike a conventional vector in three dimensions, having three components in three mutually perpendicular directions, the definition of four-vector involves concept of a four-dimensional space termed Minkowski's space. Any event in the Minkowski's four-dimensional space is represented by the coordinates (x, y, z, ict) where time has been regarded as the fourth coordinate. Thus a relativistic displacement vector must have four components rather than three components corresponding to the three dimensions of ordinary space vectors.

We consider a spherical light wave originating from the origins of the frames S and S' which coincide at the times $t = t' = 0$. The frame S', as usual, has a linear velocity v relative to frame S. The light wave in course of time spreads out in the form of a sphere. The equation of the light sphere as seen by the observer in the unprimed frame is

$$x^2 + y^2 + z^2 = c^2 t^2$$

The equation describing the light sphere in the primed frame S' is

$$x'^2 + y'^2 + z'^2 = c^2 t'^2$$

Now consider that a photon is located on the light sphere. It has both momentum and energy associated with it. The energy and the momentum carried by the photon are given by the relations

$$E = mc^2$$

$$p = mc$$

Combining these equations we get

$$p^2 = E^2/c^2 \tag{3.42}$$

If the momentum p has components p_x, p_y, p_z in frame S, and p'_x, p'_y, p'_z in frame S' we can write Eq. (3.42) for the two frames in the form

$$p_x^2 + p_y^2 + p_z^2 = E^2/c^2 \qquad (3.43)$$

$$p_x'^2 + p_{y1}'^2 + p_z'^2 = E'^2/c^2 \qquad (3.44)$$

For both sets of equations to be simultaneously true in all inertial frames, it is clear that momentum and energy must transform in the same way as spatial displacement and time. Thus the transformation equations for momentum and energy can be obtained by replacing x by p_x, y by p_y, z by p_z, and t by E/c^2 in the Lorentz transformation equations. We obtain

$$p_x' = k\,(p_x - vE/c^2); \quad p_x = k(p_x' + vE'/c^2)$$

$$p_{y,z}' = p_{y,z}; \qquad\qquad p_{y,z} = p_{y,z}' \qquad (3.45)$$

$$E' = k\,(E - vp_x); \qquad E = k\,(E' + vp_x')$$

where $\qquad k = 1/\sqrt{1 - v^2/c^2}.$

This completes the relativistic transformations for momentum and energy.

3.8. MASSLESS PARTICLES

Particles possessing zero rest mass are called massless particles. Particles belonging to this class are photons, gravitons, and neutrinos. These particles travel with the velocity of light and possess mass and energy only as long as they are in motion. When brought to rest they cease to exist. When photons, for example, are stopped by a surface they are either completely absorbed or converted to thermal energy at the surface. Since these particles travel with the velocity of light, they will be at rest with respect to a frame translating with the velocity of light. Such a frame of reference may be called the rest or the proper frame of these particles.

It can be indeed proved that particles with zero rest mass move with the velocity of light. From Eq. (3.41) we have after setting $m_0 = 0$, $E = pc$. Substituting the value of E in Eq. (3.39) yields

$$p = \frac{Eu}{c^2} = \frac{pcu}{c^2}$$

or $\qquad\qquad\qquad\qquad u = c$

which is the required result.

QUESTIONS AND PROBLEMS

3.1. Obtain the formulae for the relativistic addition of velocities according to special relativity. Show that the addition of any velocity to the velocity of light c merely reproduces the velocity of light. Also, prove that the relativistic composition of two velocities which are separately less than c, the velocity of light, can never exceed c.

3.2. Discuss the relativity of mass. Obtain a formula showing how mass varies with velocity. Describe an experiment for the verification of this formula.

3 3. What is rest mass of a particle? Can a particle have zero rest mass? Show that a particle with zero rest mass travels with the velocity of light.

3.4. What do you understand by mass-energy equivalence? Establish Einstein's mass-energy relation. How has this relation been put to experimental test?

3.5. Obtain the relativistic expression for the kinetic energy of a body and show that for very small velocities it reduces to the classical expression for kinetic energy.

3.6. What do you understand by Minkowski's space and a four-vector? Show that momentum at relativistic speeds is expressible in terms of a four-vector. Hence derive transformation equations for energy and momentum.

3.7. Show that the relativistic energy of a particle having rest mass m_0 and momentum p is given by $E = \sqrt{p^2c^2 + m_0^2 c^4}$.

3.8. Two particles approach each other with speed 2.8×10^8 m/sec with respect to the laboratory. What is their relative speed?

$$(2.993 \times 10^0 \text{ m/sec})$$

3.9. A man on the moon observes two space ships coming toward him from opposite directions at speeds of $0.8c$ and $0.9c$ respectively. What is the relative speed of the two space ships as measured by an observer in any one of them?

$$(0.988c)$$

[Hint: Here we have two velocities, $0.8c$ and $-0.9c$, as viewed by the observer on the moon. Assume an observer on the space ship moving with velocity $-0.9c$. This would correspond to a reference frame S moving with velocity $-v$ with respect to frame S'. Clearly $v = 0.9c$. Use $u' = 0.8c$

in the inverse velocity transformation

$$u = (u' + v)/1 + \frac{u'v}{c^2}]$$

3.10. A man has a mass of 100 kg on the ground. When he is in a rocket ship in flight, his mass is 101 kg as determined by an observer on the ground. What is the speed of the rocket ship?

(4.2 × 10⁷ m/sec)

3.11. The rest mass of a particle is 10 gm. What is its mass when it moves with velocity 3 × 10⁷ m/sec?

(10.09 gm)

3.12. The total energy of a particle is exactly twice its rest energy. Find its speed.

(2.598 × 10⁸ m/sec)

3.13. How many kilo-watt hours of energy must be liberated by complete conversion of 4 milli grams of mass?

(10⁵ kwH)

3.14. Calculate in electron-volt the energy corresponding to one atomic mass unit (1 amu = 1.66 × 10⁻²⁷ kg)

(931 × 10⁶ eV)

3.15. Calculate the energy liberated in MeV when a single helium nucleus is formed by the fusion of two deuterium nuclei, given

Mass of $_1H^2$ = 2.01478 amu

Mass of $_2He^4$ = 4.00388 amu

(23.91 MeV)

3.16. A certain quantity of ice at 0°C melts into water at 0°C and in the process gains 1 kg of mass. Calculate the initial mass of ice.

(2.67 × 10¹¹ kg)

[Hint: If m_0 be the initial mass of ice then the quantity of heat required to melt this mass is $m_0 \times 80$ kcal or $m_0 \times 80 \times 10^3 \times 4.2$ J. This must be equal to $(m - m_0) c^2$]

3.17. A kilogram of water is heated from 0°C to 100°C. What is increase in mass of water due to its increase in thermal energy? Could this mass increase be measured?

(4.66 × 10⁻¹² kg, no)

3.18. A photon carries energy equivalent to 10⁻¹⁹ J. Calculate the momentum of the photon in MeV/c.

(6.22 × 10⁻⁷ MeV/c)

Part Two

MECHANICS

4. Rotatory Motion

4.1. INTRODUCTION

Motion is of two types–linear and rotatory. The general motion of a rigid body is a combination of these two. We have already dealt with the linear motion in the first chapter. Rotatory motion is the motion of a body in which it rotates or revolves about a line called axis. Linear motion is caused by a force while the rotatory motion is governed by the moment of a force (torque).

In the case of linear motion, the basic variables are displacement, velocity and acceleration. In general, they are vector quantities but they may be treated as scalars when a body moves along a fixed direction. The corresponding variables involved in a rotatory motion are angular displacement, angular velocity and angular acceleration. For large angles and fixed axis of rotation they behave like scalars but for infinitesimal angular displacements and moving axis (axis not fixed) they are treated as vectors.

Angular displacement is the angle traversed by a rotating body about an axis. It may be measured in degrees but the equations of rotatory motion become much simpler and easier if it is measured in radians. By definition angular displacement θ in radians is given by the relation

$$\theta = \frac{S}{r} \tag{4.1}$$

where S is the arc length of a circle of radius r (Fig. 4.1).

If $\Delta\theta$ is the change in angular displacement of a body in Δt seconds, the instantaneous angular speed ω is defined as

$$\omega = \mathop{\mathrm{Lt}}_{\Delta t \to 0} \frac{\Delta\theta}{\Delta t} = \frac{d\theta}{dt} \tag{4.2}$$

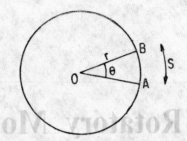

Fig. 4.1. A circle with constant radius r.

When the angular speed of a body is not constant, it has an angular acceleration. Let $\Delta\omega$ be the change in angular speed in Δt seconds. Then the instantaneous angular acceleration α is expressed as

$$\alpha = \underset{\Delta t \to 0}{\text{Lt}} \frac{\Delta\omega}{\Delta t} = \frac{d\omega}{dt} \qquad (4.3)$$

When the axis of rotation is fixed, the angular speed ω and the angular acceleration α of a rigid body rotating about this axis remain constant, as all the radial lines fixed in the rigid body perpendicular to the axis of rotation rotate through the same angle in the same time. Thus ω and α are the characteristic of the body as a whole. The dimensions of ω and α are T^{-1} and T^{-2} respectively.

4.2. RELATION BETWEEN LINEAR AND ANGULAR PARAMETERS FOR A PARTICLE IN CIRCULAR MOTION

Let P be a particle moving in a circle of constant radius r (Fig. 4.2). The position vector $\vec{r}$ of P at any time t is given by

$$\vec{r} = r\hat{r} = x\hat{i} + y\hat{j}$$

where $\hat{r}$, $\hat{i}$ and $\hat{j}$ are the unit vectors along $\vec{r}$, x-axis and y-axis respectively. If ω is the angular speed of P, then $\theta = \omega t$. Hence we can write

$$\vec{r} = r\cos\omega t\,\hat{i} + r\sin\omega t\,\hat{j}$$

or $\qquad\qquad\qquad \vec{r} = r(\cos\omega t\,\hat{i} + \sin\omega t\,\hat{j}) \qquad (4.4)$

The velocity $\vec{V}$ of particle P (along the tangent) is obtained by differentiating Eq. (4.4) with respect to time; that is,

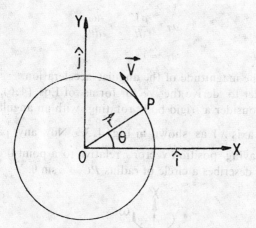

Fig. 4.2. A particle moving in a circle.

$$\vec{V} = \frac{d\vec{r}}{dt} = r\omega(-\sin \omega t \,\hat{i} + \cos \omega t \,\hat{j}) \qquad (4.5)$$

Therefore, the magnitude of $\vec{V}$ is given by

$$V = \sqrt{\omega^2 r^2(\sin^2 \omega t + \cos^2 \omega t)}$$

or $$V = \omega r \qquad (4.6)$$

Eq. (4.6) is the scalar relation between V and ω.

Differentiating Eq. (4.5) with respect to time, we get

$$\vec{a}_R = \frac{d\vec{V}}{dt} = -\omega^2 r(\cos \omega t \,\hat{i} + \sin \omega t \,\hat{j})$$

or $$\vec{a}_R = -\omega^2 \vec{r} \quad \text{(from Eq. 4.4)}$$

or $$\vec{a}_R = -\omega^2 r \,\hat{r} \qquad (4.7)$$

where $\vec{a}_R$ is the radial component of the acceleration or the centripetal acceleration. The negative sign indicates that $\vec{a}_R$ is always directed towards the centre of the circle. The magnitude of $\vec{a}_R$ is written as

$$a_R = \omega^2 r = \omega v \quad (\because v = r\omega) \qquad (4.8)$$

In order to obtain the tangential component of the acceleration, we differentiate Eq. (4.6) with respect to time; that is,

$$a_T = \frac{dV}{dt} = r\frac{d\omega}{dt}$$

or $$a_T = r\alpha \tag{4.9}$$

where α is the magnitude of the angular acceleration.

In order to derive the vector forms of Eqs. (4.6), (4.8) and (4.9), we consider a rigid body rotating with an angular velocity $\vec{\omega}$ about an axis XY as shown in Fig. 4.3. Now any particle P of the body, having position vector $\vec{r}$ relative to a point 0 on the axis of rotation, describes a circle of radius $PC = r\sin\theta$.

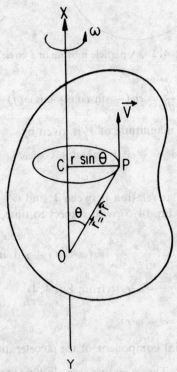

Fig. 4.3.   A rigid body rotating about an axis XY.

Using Eq. (4.6), the magnitude of the linear velocity of the particle P is written as

$$V = \omega r \sin\theta$$

or $$\vec{V} = \vec{\omega} \times \vec{r} \tag{4.10}$$

Eq. (4.10) gives the vector relation between $\vec{V}$ and $\vec{\omega}$.

Differentiating Eq. (4.10) with respect to time, we get

$$\frac{d\vec{V}}{dt} = \vec{\omega} \times \frac{d\vec{r}}{dt} + \frac{d\vec{\omega}}{dt} \times \vec{r}$$

or

$$\vec{a} = \vec{\omega} \times \vec{V} + \vec{\alpha} \times \vec{r} \qquad (4.11)$$

where $\vec{a}$ is the resultant acceleration of the particle P. Comparing Eq. (4.11), with Eqs. (4.8) and (4.9), we conclude

$$\vec{a}_R = \vec{\omega} \times \vec{V}, \quad \vec{a}_T = \vec{\alpha} \times \vec{r} \qquad (4.12)$$

Eq. (4.12) gives the vector form of the radial and tangential accelerations.

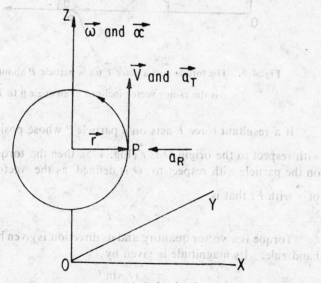

Fig. 4.4. The directions of vectors $\vec{\omega}$, $\vec{\alpha}$, $\vec{a}_T$, $\vec{a}_R$, $\vec{V}$ and $\vec{r}$ for a particle rotating in a circle about z-axis.

The directions of $\vec{\omega}$, $\vec{\alpha}$, $\vec{a}_T$, $\vec{a}_R$, $\vec{V}$ and $\vec{r}$ are shown in Fig. 4.4.

The directions of angular parameters $\vec{\omega}$ and $\vec{\alpha}$ are given by right hand rule; that is, *"If the right hand is held so that curled fingers follow the rotation of P, the extended right thumb will point in the direction*

$of \vec{\omega}$ *and* $\vec{\alpha}$. In Fig. 4.4 we find that $\vec{\omega}$ is at right angles to $\vec{V}$, and $\vec{\alpha}$ is at right angles to $\vec{r}$. Hence Eq. (4.12) gives

$$a_R = \omega v \text{ and } a_T = \alpha r$$

which are the same as in Eqs. (4.8) and (4.9) respectively.

4.3. TORQUE AND ANGULAR MOMENTUM

For a single particle, the torque is defined as the moment of the resultant force on the particle with respect to a particular origin.

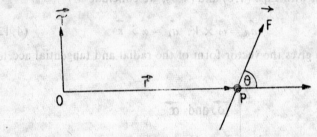

Fig. 4.5. The torque $\vec{\tau}$ of a force $\vec{F}$ on a particle P about O. $\vec{r}$ is the radius vector inclined at an angle θ to $\vec{F}$.

If a resultant force $\vec{F}$ acts on a particle P whose position vector with respect to the origin O is $\vec{r}$ (Fig. 4.5), then the torque $\vec{\tau}$ acting on the particle with respect to O is defined as the vector product of $\vec{r}$ with $\vec{F}$; that is,

$$\vec{\tau} = \vec{r} \times \vec{F} \qquad (4.13)$$

Torque is a vector quantity and its direction is given by the right hand rule. Its magnitude is given by

$$\tau = rF \sin \theta \qquad (4.14)$$

where θ is the angle between $\vec{r}$ and $\vec{F}$. From Eq. (4.14), τ can be defined as the moment of the force F about O. As the rotatory motion is caused by the moment of a force, τ can be called the rotational analog of the force in the linear motion. Dimensions of torque are the same as those of work but torque and work are two different physical quantities. Torque is a vector while the work is a scalar quantity.

Angular momentum of a particle is defined as the moment of its linear momentum. Mathematically,

$$\vec{J} = \vec{r} \times \vec{p} \qquad (4.15)$$

where $\vec{J}$ is the angular momentum of a particle P of mass m whose linear momentum is $\vec{p}$ and position vector with respect to origin O of an inertial frame (XYZ, O) is $\vec{r}$ as shown in Fig. 4.6.

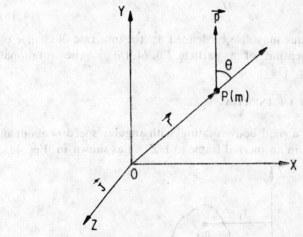

Fig 4.6. The direction of angular momentum of a particle P about O. The radius vector $\vec{r}$ and $\vec{p}$ (momentum) are inclined at an angle θ.

Angular momentum is a vector quantity and its direction (along z-axis) is normal to the plane (XY) formed by $\vec{r}$ and $\vec{p}$. Its magnitude is given by

$$J = rp \sin \theta \qquad (4.16)$$

where θ is the angle between $\vec{r}$ and $\vec{p}$.

Differentiating Eq. (4.15) with respect to time, we have

$$\frac{d\vec{J}}{dt} = \frac{d\vec{r}}{dt} \times \vec{p} + \vec{r} \times \frac{d\vec{p}}{dt}$$

or $$\frac{d\vec{J}}{dt} = \vec{V} \times (m\vec{V}) + \vec{r} \times \vec{F} \qquad (4.17)$$

Here $$\vec{V}\left(=\frac{dr}{dt}\right) \text{ and } \vec{F}\left(=\frac{dp}{dt}\right)$$

are, respectively, the velocity and force acting on the particle P. Eq. (4.17) can further be simplified to

$$\frac{d\vec{J}}{dt} = \vec{r} \times \vec{F} \qquad\qquad [\because \quad \vec{V} \times (m\vec{V}) = 0]$$

Hence $$\vec{\tau} = \frac{d\vec{J}}{dt} \qquad\qquad\qquad (4.18)$$

Thus torque may also be defined as the time rate of change of angular momentum of a particle. Eq. (4.18) is the rotational analog of Eq. (1.5).

4.4. MOMENT OF INERTIA

Consider a rigid body rotating with angular speed ω about an axis OZ fixed in an inertial frame (XYZ, O) as shown in Fig. 4.7.

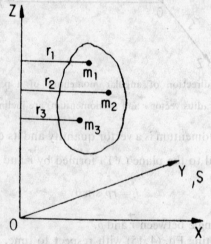

Fig. 4.7. A rotating rigid body.

We assume the rigid body to be composed of large number of particles of masses $m_1, m_2, m_3 \ldots$ etc. situated at distances $r_1, r_2, r_3, \ldots$ respectively from the axis of rotation. If $v_1, v_2, v_3, \ldots$ etc. are, respectively, the linear velocities of $m_1, m_2, m_3, \ldots$, the total kinetic energy of the rotating body is written as

$$K = \tfrac{1}{2}m_1 v_1{}^2 + \tfrac{1}{2}m_2 v_2{}^2 + \ldots \qquad (4.19)$$

Using Eq. (4.6) into Eq. (4.19), we get

$$K = \tfrac{1}{2}m_1 r_1{}^2 \omega^2 + \tfrac{1}{2}m_2 r_2{}^2 \omega^2 + \ldots$$

$$= \tfrac{1}{2}(m_1 r_1{}^2 + m_2 r_2{}^2 + \ldots) \omega^2 \quad [\because \ \omega = \text{const.}]$$

or $$K = \tfrac{1}{2}\omega^2 \sum_{i=1}^{n} m_i r_i{}^2$$

or $$K = \tfrac{1}{2}I\omega^2 \qquad (4.20)$$

where $$I \left(= \sum_{i=1}^{n} m_i r_i{}^2 \right)$$

is called the moment of inertia or the rotational inertia of the rotating body about the given axis of rotation.

Hence the moment of inertia of a rigid body about a given axis of rotation is defined as the sum of the products of the masses of its particles and the squares of their respective distances from the axis of rotation.

Note that the moment of inertia of a body depends on (a) the axis of rotation and (b) the manner in which mass of the body is distributed about the given axis.

Eq. (4.20) is analogous to the expression for linear kinetic energy ($\tfrac{1}{2}Mv^2$) of a body. As v is analogous to ω, the moment of inertia I plays the same role as mass M does in the linear motion. But the mass of a body remains constant while the moment of inertia of a body can be altered by varying the axis of rotation and the distribution of mass about the axis.

From analogy between linear and rotatory motion, Eq. (1.8) of linear motion transforms to

$$\vec{\tau} = I\vec{\alpha} \qquad (4.21)$$

where $\vec{\tau}$ is the torque and $\vec{\alpha}$ is the angular acceleration.

From Eqs. (4.18) and (4.21), we get

$$\frac{d\vec{J}}{dt} = I\vec{\alpha}$$

or in the scalar form

$$\frac{dJ}{dt} = I\alpha$$

or
$$\frac{dJ}{dt} = I\frac{d\omega}{dt}$$

or
$$\frac{dJ}{dt} = \frac{d}{dt}(I\omega)$$

$$(\because I = \text{const. for a fixed axis})$$

Hence $J = I\omega$ (4.22)

Eq. (4.22) is analogous to $p = mv$ in the linear motion.

Therefore, if we compare the linear and rotatory motions of a body, we find that for every physical quantity in the linear motion there is an analogous quantity in the rotatory motion. Table 4.1 gives the comparison between the various physical quantities for a linear motion (along a fixed straight line) and the rotational motion of a rigid body about a fixed axis.

Table 4.1

Rectilinear Motion		Rotation about a fixed axis	
Displacement	r	Angular displacement	θ
Velocity	$v = \dfrac{dr}{dt}$	Angular velocity	$\omega = \dfrac{d\theta}{dt}$
Acceleration	$a = \dfrac{dv}{dt}$	Angular acceleration	$\alpha = \dfrac{d\omega}{dt}$
Mass	M	Moment of Inertia	I
Force	$F = Ma$	Torque	$\tau = I\alpha$
Work	$W = \int F dr$	Work	$\int \tau d\theta$
Kinetic Energy	$\frac{1}{2} Mv^2$	Kinetic Energy	$\frac{1}{2} I\omega^2$
Power	$P = Fv$	Power	$P = \tau\omega$
Linear Momentum	$p = Mv$	Angular Momentum	$J = I\omega$

Radius of Gyration

If the mass of a rotating body is assumed to be concentrated at one point situated at a distance K from the axis of rotation, K is known as the radius of gyration of the body about the axis of rotation. In such a case the moment of inertia of the body about the given axis of rotation is expressed as

$$I = MK^2 \qquad (4.23)$$

where M is the total mass of the body.

4.5. THEOREMS OF MOMENT OF INERTIA

There are two important theorems of moment of inertia. They are known as theorems of parallel and perpendicular axes. Let us discuss them one by one.

1. Theorem of parallel axes states that *the moment of inertia of a body about a given axis is equal to the sum of its moment of inertia about a parallel axis through its centre of gravity and the product of the mass of the body with the square of the perpendicular distance between the two axes.* Mathematically

$$I = I_G + Ma^2 \qquad (4.24)$$

where I is the moment of inertia of the body about the given axis, I_G is the moment of inertia about a parallel axis through its centre of gravity, M is the mass of the body and a is the perpendicular distance between the two axes.

Proof. Let XY be the given axis of rotation of a rigid body (Fig. 4.8). AB is a parallel axis through the centre of gravity of the body. A particle P of mass m (of the body) is situated at a distance x from AB. a is the perpendicular distance between XY and AB.

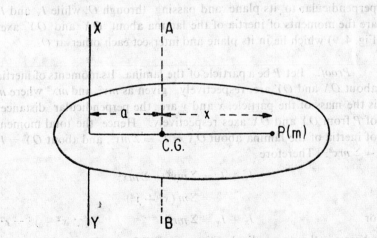

Fig. 4.8. Two parallel axes of rotation.

By definition the moment of inertia of P about

$$XY = m(x + a)^2$$

and $\qquad\qquad AB = mx^2$

Hence the moment of inertia of the whole body, I, about XY is given by

$$I = \Sigma m (x + a)^2$$
$$= \Sigma mx^2 + \Sigma ma^2 + 2\Sigma max$$

or $\qquad\qquad I = I_G + Ma^2 + 2a \Sigma mx$

Now Σmx is the sum of the moments of all the particles about the centre of gravity.

Hence $\qquad\qquad \Sigma mx = 0$

Therefore $\qquad\qquad I = I_G + Ma^2$

2. **Theorem of Perpendicular axes** states that *the moment of inertia of a plane lamina about an axis perpendicular to its plane is equal to the sum of its moments of inertia about two axes mutually perpendicular, in the plane of the lamina, and intersecting each other at a point through which the perpendicular axis passes.* Mathematically,

$$I = I_x + I_y \qquad\qquad (4.25)$$

where I is the moment of inertia of a plane lamina about an axis perpendicular to its plane and passing through O, while I_x and I_y are the moments of inertia of the lamina about OX and OY axes Fig. 4.9) which lie in its plane and intersect each other at O.

Proof. Let P be a particle of the lamina. Its moments of inertia about OX and OY are, respectively, given as my^2 and mx^2 where m is the mass of the particle, x and y are the perpendicular distances of P from OY and OX axes respectively. Hence the total moment of inertia of the lamina about $OX = I_x = \Sigma my^2$, and about $OY = I_y = \Sigma mx^2$. Therefore

$$I_x + I_y = \Sigma my^2 + \Sigma mx^2$$
$$= \Sigma m (x^2 + y^2)$$

or $\qquad\qquad I_x + I_y = \Sigma mr^2 \qquad\qquad (\because x^2 + y^2 = r^2)$

where r is the perpendicular distance of P from a perpendicular axis

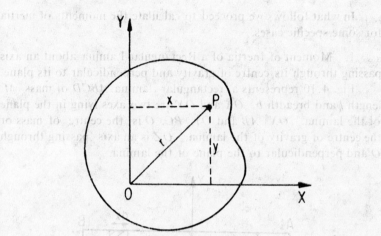

Fig. 4.9. Perpendicular axes.

passing through O. But by definition the moment of inertia of P about a perpendicular axis through $O = \Sigma\, mr^2 = I\,(\text{say})$

Hence $\qquad\qquad\qquad I = I_x + I_y$

4.6. CALCULATION OF MOMENT OF INERTIA

To calculate the moment of inertia of a continuous, homogeneous body having a definite geometrical shape, we proceed as under:

(i) We consider a small element of the body and calculate its moment of inertia about the given axis by using the basic definition of moment of inertia; that is,

$$I_E = dm\, x^2 \qquad (4.26)$$

where I_E is the moment of inertia of a small element of mass dm situated at a distance x from the axis of rotation.

(ii) We calculate the moment of inertia of the whole body by integrating Eq. (4.26) over the limits depending on the shape of the body; that is,

$$I = \int I_E = \int dm\, x^2$$

(iii) Wherever necessary, we make use of the theorems of parallel and perpendicular axes.

In what follows we proceed to calculate the moments of inertia for some specific cases.

1. Moment of Inertia of a Rectangular Lamina about an axis passing through its centre of gravity and perpendicular to its plane.

Fig. 4.10 represents a rectangular lamina $ABCD$ of mass M, length l and breadth b. OX and OY are two axes lying in the plane of the lamina. $OX \parallel AB$ and $OY \parallel BC$. O is the centre of mass or the centre of gravity of the lamina. OZ is an axis passing through O and perpendicular to the plane of the lamina.

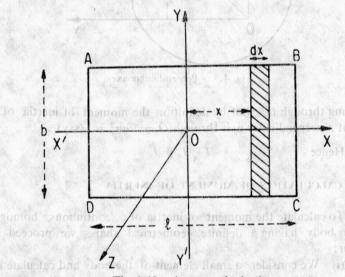

Fig. 4.10.   A rectangular lamina.

Consider a small element dx of the lamina parallel to OY and situated at a distance x from it.

Mass per unit length of the lamina $= M/l$. Therefore the mass of the element $dx = (M/l)\,dx$.

Now moment of inertia of element dx about

$$OY = (M/l)\,dx\,x^2$$

Therefore, the moment of inertia of the whole lamina about OY is given by

$$I_y = \int_{-l/2}^{l/2} (M/l)\,x^2\,dx$$

$$= \frac{M}{l}\left[\frac{x^3}{3}\right]_{-l/2}^{l/2} = \frac{M}{l}\left[\frac{l^3}{24} + \frac{l^3}{24}\right]$$

or
$$I_y = \frac{Ml^2}{12} \tag{4.27}$$

Following the procedure adopted for the determination of I_y, we can prove that moment of inertia of the lamina about OX axis is given by

$$I_x = \frac{Mb^2}{12} \tag{4.28}$$

Using the theorem of perpendicular axes, the moment of inertia of the lamina about OZ axis is written as

$$I_z = I_x + I_y$$

or
$$I_z = M\frac{(l^2 + b^2)}{12} \tag{4.29}$$

2. **Moment of Inertia of a Solid Uniform Bar of Rectangular Cross-Section about an axis passing through its centre of mass and perpendicular to its length.** In Fig. 4.11, $ABCDEFGH$ is a rectangular bar of mass M, length l, breadth b and depth d. YOY' is an axis passing through the centre of mass of the bar and perpendi-

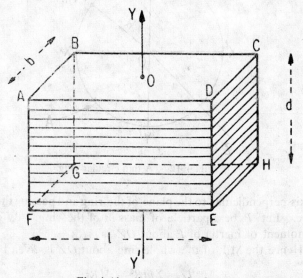

Fig. 4.11. A rectangular bar.

cular to its length.   Divide the rectangular bar into a number of
plane laminas placed one above the other as shown in Fig. 4.11.

If m, l and b are the mass, length and breadth of each lamina
respectively, the moment of inertia of each lamina is given by Eq.
(4.29); that is, $m\left(\dfrac{l^2 + b^2}{12}\right)$. Hence the moment of inertia of the
rectangular bar is the sum of the moments of inertia of all laminas
taken together.   Thus

$$I = \Sigma m \left(\frac{l^2 + b^2}{12}\right)$$

$$= \frac{l^2 + b^2}{12} \Sigma m$$

or
$$I = M\frac{(l^2 + b^2)}{12} \tag{4.30}$$

3.   Moment of Inertia of a Thin Circular Ring. (i) About an
axis through its centre and perpendicular to its plane.

A is a thin circular ring of radius R and centre O. XOX'
and YOY' are two axes in the plane of the ring (Fig. 4.12). OZ is

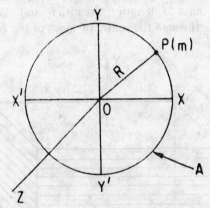

Fig. 4.12.   A thin circular ring.

an axis perpendicular to the plane of the ring and passing through its
centre.   Let P be a particle of mass m of the ring.   By definition
the moment of inertia of P about OZ is mR^2.

Hence the M.I. of the whole ring about OZ is given by

$$I_z = \Sigma m R^2 = M R^2 \tag{4.31}$$

(ii) About a diameter.

Using the theorem of perpendicular axes,

$$I_Z = I_x + I_y$$

where I_x and I_y are the moments of inertia of the ring about XOX' and YOY' respectively.

Since the ring is symmetrical about its diameter $I_x = I_y = I$ (say),

Hence $$2I = I_Z = MR^2$$

or $$I = \frac{MR^2}{2} \qquad (4.32)$$

4. Moment of Inertia of a Circular Disc. (i) About an axis passing through its centre and normal to its plane.

Fig. 4.13 represents a disc A of mass M, radius R and centre O. Consider a small circular element dx of the disc at a distance x from O.

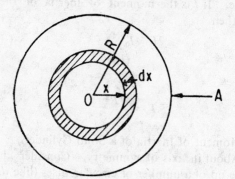

Fig. 4.13. A circular disc of radius R.

Now mass per unit area of the disc $= \dfrac{M}{\pi R^2}$

Hence mass of the small element

$$dx = \frac{M}{\pi R^2}\, 2\pi x\, dx = \frac{2M}{R^2}\, x\, dx$$

Using Eq. (4.31), the moment of inertia of the element dx about an axis passing through O and normal to the plane of the disc is written as

$$I_E = \frac{2M}{R^2}\, x dx \cdot x^2$$

or $$I_E = \frac{2M}{R^2} x^3 dx \qquad (4.33)$$

Hence the moment of inertia of the entire disc about the said axis is obtained by integrating Eq. (4.33) between the limits $x = 0$ and $x = R$; that is,

$$I_0 = \int_0^R \frac{2M}{R^2} x^3 dx$$

$$= \frac{2M}{R^2} \left[\frac{x^4}{4} \right]_0^R = \frac{2M}{R^2} \cdot \frac{R^4}{4}$$

or $$I_0 = \frac{MR^2}{2} \qquad (4.34)$$

(ii) About a diameter. Since the disc is symmetrical about any of its diameters, its moment of inertia about every diameter is the same. If I is the moment of inertia of the disc about a diameter, then

$$2I = I_0 \qquad\qquad (\perp \text{ axes theorem})$$

$$= \frac{MR^2}{2}$$

or $$I = \frac{MR^2}{4} \qquad (4.35)$$

5.   Moment of Inertia of a Solid Cylinder.
(i)   About its axis of symmetry. Consider the solid cylinder to be made up of a number of circular discs (like dx, Fig. 4.14) of mass m each. If R is the radius of the cylinder, then the moment of inertia of each disc about XOX' (axis of symmetry) is $mR^2/2$ (Eq. 4.34).
Hence the moment of inertia of the cylinder about its axis of symmetry is given by

$$I = \Sigma \frac{mR^2}{2} = \frac{R^2}{2} \Sigma m$$

or $$I = \frac{MR^2}{2} \qquad (4.36)$$

where M ($= \Sigma m$) is the mass of the cylinder.
(ii)   About an axis normal to its axis of symmetry and passing through its centre of mass.

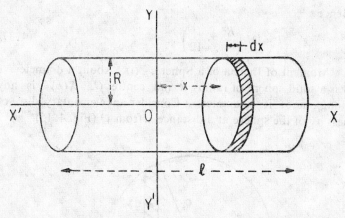

Fig. 4.14. A solid cylinder of radius R.

We consider an elementary disc dx of the cylinder situated at a distance x from the given axis (YOY', Fig. 4.14). Let l be the length of the cylinder. Now mass per unit length of the cylinder is M/l.

Hence mass of the elementary disc $= (M/l)\, dx$. Moment of inertia of elementary disc about a diameter parallel to YOY' is written as

$$I_E = \frac{M}{l}\, dx \cdot \frac{R^2}{4} \quad \text{(see Eq. 4.35)}$$

Using parallel axes theorem the moment of inertia of elementary disc about

$$YOY' = \frac{M}{l} dx \frac{R^2}{4} + \frac{M}{l} dx\, x^2$$

Therefore, the moment of inertia of the entire cylinder about YOY' (an axis passing through its centre of mass O) is expressed as

$$I = \int_{-l/2}^{l/2} \left[\frac{MR^2}{4l}\, dx + \frac{M}{l} x^2\, dx \right]$$

$$= \frac{MR^2}{4l} \left[x \Big[_{-l/2}^{l/2} + \frac{M}{l} \Big[\frac{x^3}{3} \Big]_{-l/2}^{l/2} \right.$$

or
$$I = \frac{MR^2}{4} + \frac{Ml^2}{12}$$

Hence

$$I = M\left(\frac{R^2}{4} + \frac{l^2}{12}\right) \tag{4.37}$$

6. Moment of Inertia of a Sphere.   (i) About a diameter.

A is a solid sphere of radius *R* and centre *O*. *XOX'* is any axis along one of its diameters. Consider an elementary disc of thickness *dx* of the sphere at a distance *x* from *O* (Fig. 4.15).

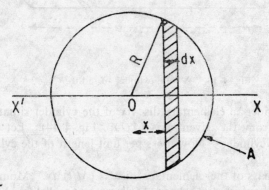

Fig. 4.15.   A solid sphere of radius *R*.

Let ρ be the mass per unit volume of the sphere.

Now radius of the disc $= (R^2 - x^2)^{1/2}$. Therefore volume of the disc $= \pi (R^2 - x^2)\, dx$. Hence mass of the

$$\text{disc} = \rho\pi\,(R^2 - x^2)\, dx$$

Using Eq. (4.34), the moment of inertia of the disc about *XOX'* is written as

$$I_d = \frac{\rho\pi(R^2 - x^2)\, dx\,(R^2 - x^2)}{2}$$

or $$I_d = \frac{\rho\pi\,(R^2 - x^2)^2}{2}\, dx$$

Hence the moment of inertia of the sphere about *XOX'* is given by

$$I = \int\limits_{-R}^{+R} \frac{\rho\pi\,(R^2 - x^2)^2}{2}\, dx$$

$$= \frac{\rho\pi}{2} \int\limits_{-R}^{+R} (R^4 + x^4 - 2x^2R^2)\, dx$$

$$= \frac{\rho\pi}{2} \left[R^4 x + \frac{x^5}{5} - \frac{2R^2 x^3}{3} \right]_{-R}^{+R}$$

$$= \frac{\rho\pi}{2} \left[2R^5 + \frac{2R^5}{5} - \frac{4R^5}{3} \right]$$

$$= \frac{8\pi\rho R^5}{15} = \frac{4}{3} \pi R^3 \rho \times \frac{2R^2}{5}$$

or $\qquad\qquad I = \frac{2}{5} MR^2 \qquad\qquad\qquad$ (4.38)

where $M \; (= \frac{4}{3} \pi R^3 \rho)$ is the mass of the sphere.

(ii) About a tangent.

Any tangent to the sphere is parallel to one of its diameters and is situated at a distance equal to the radius R of the sphere.

Thus using the theorem of parallel axes, the moment of inertia of the sphere about any of its tangents is given by

$$I_t = 2/5\, MR^2 + MR^2$$

or $\qquad\qquad I_t = 7/5\, MR^2. \qquad\qquad\qquad$ (4.39)

4.7. CONSERVATION OF ANGULAR MOMENTUM

The law of conservation of angular momentum states that the total angular momentum of a system remains constant in the absence of external torques. Since the angular momentum is a vector quantity, the total angular momentum of a system of n particles in a given reference frame is given by the vector sum of the angular momenta of individual particles in the same frame; that is,

$$\vec{J} = \vec{J_1} + \vec{J_2} + \vec{J_3} + \dots \vec{J_n} \qquad\qquad (4.40)$$

where $\vec{J}$ is the total angular momentum and $\vec{J_1}, \vec{J_2}, \vec{J_3}$ etc. are the angular momenta of individual particles.

Using Eqs. (4.18) and (4.40), the total torque acting on the system is given by

$$\vec{\tau} = \frac{d\vec{J}}{dt} = \frac{d}{dt} (\vec{J_1} + \vec{J_2} + \dots \vec{J_n}) \qquad\qquad (4.41)$$

Now the total torque $\vec{\tau}$ acting on the system is the sum of external and internal torques. Thus

$$\vec{\tau} = \vec{\tau}_{ext} + \vec{\tau}_{int} \qquad (4.42)$$

where $\vec{\tau}_{ext}$ is the torque due to external forces and $\vec{\tau}_{int}$ (internal torque) is due to the moments of the mutual interaction of the particles on one another. The internal forces of action and reaction exist in pairs and are equal in magnitude and opposite in direction, therefore their moments about a point annul each other.

Hence $\qquad\qquad\qquad \vec{\tau}_{int} = 0 \qquad\qquad\qquad (4.43)$

Using Eqs. (4.43) and (4 42) in Eq. (4.41), we get

$$\vec{\tau}_{ext} = \frac{d\vec{J}}{dt} = \frac{d}{dt}(\vec{J}_1 + \vec{J}_2 + \ldots \vec{J}_n) \qquad (4.44)$$

In the absence of external torque ($\vec{\tau}_{ext} = 0$) Eq. (4.44) yields

$$\vec{J} = \vec{J}_1 + \vec{J}_2 + \ldots \vec{J}_n = \text{constant} \qquad (4.45)$$

Eq. (4.45) is the mathematical statement of the law of conservation of angular momentum.

For a fixed axis of rotation, the angular momentum of a rigid body is written as

$$J = I\omega \quad \text{(see Eq. 4.22)}$$

where I is the moment of inertia and ω is the angular velocity of the rotating body.

Hence, for this case, the principle of angular momentum can also be expressed as

$$I\omega = \text{constant} \qquad (4.46)$$

4.8. SOME EXAMPLES OF CONSERVATION OF ANGULAR MOMENTUM

1. This principle is used by acrobats, divers, ballet dancers, ice skaters and others to increase or decrease their angular speeds about the axis of rotation. This is done by changing the moment of inertia of the rotating body by rearrangement of its parts about

the axis of rotation. If the moment of inertia changes, there must be a corresponding change in ω so that equation (4.46) is satisfied.

2. *Motion of Satellite and Planets.* Planets move round the sun under the action of central forces (gravitational forces). A central force acting on a particle is one which is always directed towards or away from a point and whose magnitude depends on the distance r of the particle from that point. Mathematically,

$$\vec{F} = f(r)\,\hat{r}$$

where $\vec{F}$ is a central force, $f(r)$ is a scalar function of r and $\hat{r}\left(=\dfrac{\vec{r}}{r}\right)$ is a unit vector along r.

Now

$$\vec{\tau} = \frac{d\vec{J}}{dt} = \vec{r} \times \vec{F}$$

or

$$\frac{d\vec{J}}{dt} = \vec{r} \times f(r)\,\hat{r}$$

$$= \vec{r} \times f(r)\frac{\vec{r}}{r}$$

$$= \frac{f(r)}{r}(\vec{r} \times \vec{r}) = 0$$

Hence

$$\frac{d\vec{J}}{dt} = 0$$

or

$$\vec{J} = \text{constant}$$

Thus the angular momentum of a body under the action of a central force is constant. This means that the angular momentum of satellites and planets is always conserved and also their paths are situated in a plane. The angular momentum of a planet of mass m and moving with a velocity v at a distance r from the sun is mvr. If r is less, v has to be more so that the angular momentum (mvr) of the planet remains constant. Therefore, a planet will move faster at a point closer to the sun than at a point farther from it.

Kepler's second law of planetary motion, which states that the areal velocity of a planet round the sun is constant, is also an outcome of the law of conservation of angular momentum.

3. *Shape of galaxy.* A galaxy is a group of innumerable stars consisting of huge amounts of condensed gases. The shape of a galaxy is like that of a convex lens. It is thought that galaxies have been formed by the gravitational condensation of huge collection of gas molecules. Let us assume that, initially, the galaxy was a collection of gas molecules possessing angular momentum and contracting due to its own gravitational field. (Till the present day it is not known where the gases came from and why an angular momentum should be assigned to them. We only know that masses without angular momentum will condense as spheres.) The contraction of gas molecules will continue both parallel and perpendicular to the direction of angular momentum till the mass acquires a minimum radius. Beyond this a contraction in a direction perpendicular to the angular momentum which is conserved, is not possible as the gravitational energy is not strong enough to cause the contraction. However, a contraction parallel to the angular momentum is still possible as this does not involve any energy change. The present shape (Fig. 4.16b) of the galaxy is probably due to the continued contraction of gases parallel to the angular momentum $\vec{J}$. Fig. 4.16a shows the initial shape of the galaxy.

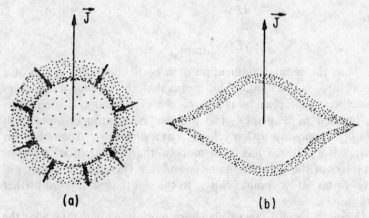

(a)   (b)

Fig. 4.16.   Shape of galaxy.

4. *Scattering of charged particles (protons or α-particles) by a heavy nucleus.* Let us consider a proton of mass m and charge e approaching a heavy nucleus A containing charge Ze, where Z is the atomic number. As the proton moves near the nucleus it will experience an electrostatic force of repulsion and will follow a hyperbolic path as shown in Fig. 4.17.

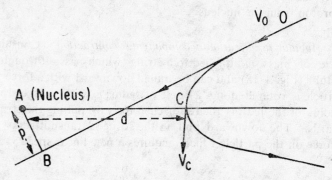

Fig. 4.17.   Scattering of a proton by a nucleus.

At O, a distant point, the velocity of the proton is v_o and at C, the closest point, its velocity is v_c. d is the distance of closest approach and p is the impact parameter (perpendicular distance of A from the direction of proton velocity at O).

Now angular momentum of the proton at

$$O = mv_o p$$

and its angular momentum at

$$C = mv_c d$$

Energy of the proton at $O = \tfrac{1}{2}mv_o^2$

Energy of the proton at $C = \tfrac{1}{2}mv_c^2 + \dfrac{Ze^2}{d}$

Applying the laws of conservation of angular momentum and energy, we can write

$$mv_c d = mv_o p \qquad\qquad (4.47)$$

and
$$\tfrac{1}{2}mv_o^2 = \tfrac{1}{2}mv_c^2 + \dfrac{Ze^2}{d} \qquad\qquad (4.48)$$

Substituting the value of v_c from Eq. (4.47) into Eq. (4.48), we have

$$\tfrac{1}{2}mv_o^2 = \tfrac{1}{2}m\left(\frac{v_o\,p}{d}\right)^2 + \frac{Ze^2}{d}$$

or
$$\frac{Ze^2}{d} = \tfrac{1}{2}mv_o^2\left[1 - \left(\frac{p}{d}\right)^2\right]. \tag{4.49}$$

Eq. (4.49) can be used to calculate the distance of closest approach of the proton from the nucleus.

5. *Angular acceleration accompanying contraction.* Consider a particle of mass m attached to a string which passes through a hollow tube (Fig. 4.18) and can be pulled downward with a force F. The particle is rotated along a circle of radius r_1 with a speed v_1. The radius of the path is then shortened to r_2 by pulling the string downwards. The downward pull on the string is transmitted as a radial force on the particle which acquires a new linear speed v_1.

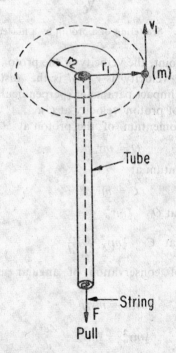

Fig. 4.18. A mass at the end of a string moves in a circle of radius r_1 with angular speed ω_1. The string passes down through a tube. F supplies the centripetal force.

Since the particle is acted upon by a central force, its angular momentum is conserved. Hence

Initial angular momentum = Final angular momentum;
that is,

$$mv_1r_1 = mv_2r_2$$

or
$$mr_1^2\omega_1 = mr_2^2\omega_2 \quad (\because v = r\omega)$$

or
$$\omega_2 = \left(\frac{r_1}{r_2}\right)^2 \omega_1 \qquad (4.50)$$

where ω_1 and ω_2 are the initial and final angular speeds of the particle respectively.

Since $r_1 > r_2$ the particle will move with greater angular speed on being pulled in. Note that the contraction of the string is followed by an angular acceleration of the particle.

4.9. ROTATIONAL SYMMETRY

Rotational symmetry means that the laws of physics remain the same during rotation through a fixed angle. That means the angular orientation has no effect on the physical laws.

Let us consider two observers A and B stationed in coordinate systems (XY, O) and $(X'Y', O)$ respectively. To simplify the problem the two coordinate systems are restricted to two dimensions and the observers A and B use the common origin. The axes of A are rotated through an angle θ relative to those of B as shown in Fig. 4.19a. P is any point having coordinates (x, y) in A's system and (x', y') in B's system. The coordinates of P in the two systems are related as

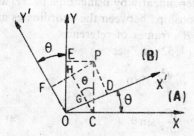

Fig. 4.19 a. Two coordinate systems having different angular orientations.

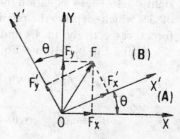

Fig. 4.19 b. Components of a force in the two systems.

$$x' = OG + GD = OG + PH$$

or
$$x' = x\cos\theta + y\sin\theta \qquad (4.51)$$

$$y' = PD = HC - GC = y\cos\theta - x\sin\theta$$

$$z' = z$$

Fig. 4.19b is used to analyse the relationship of forces as seen by the two observers. Let us assume that a force F, whose components in A's frame are F_x, F_y and F_z, is acting on a particle of mass m located at a point O (Fig. 4.19b). For simplicity we shift the origin of the coordinate systems to P (as we can have linear displacements without disturbing the physical laws). Now the components of F as seen by B along his axes are F_x', F_y' and F_z'. Using the Fig. 4.19b, these components can be expressed in terms of F_x, F_y and F_z. The results are

$$F_x' = F_x\cos\theta + F_y\sin\theta$$

$$F_y' = F_y\cos\theta - F_x\sin\theta \qquad (4.52)$$

and
$$F_z' = F_z$$

If we assume Newtou's laws to be true in A's system, then we can express them as

$$F_x = m\frac{d^2x}{dt^2}, \quad F_y = m\frac{d^2y}{dt^2}, \quad F_z = m\frac{d^2z}{dt^2} \qquad (4.53)$$

If B, in his rotated frame, can apply Newton's laws, then we can write

$$F_x' = m\frac{d^2x'}{dt^2}, \quad F_y' = m\frac{d^2y'}{dt^2}, \quad F_z' = m\frac{d^2z'}{dt^2} \qquad (4.54)$$

To test the validity of Eq. (4.54), we evaluate the left and right sides of this equation independently by using Eqs. (4.51) and (4.52) which express the relationship between the coordinates and forces, respectively, in A's and B's frames of reference.

Using Eq. (4.53) into Eq. (4.52), we get

$$F_x' = m\frac{d^2x}{dt^2}\cos\theta + m\frac{d^2y}{dt^2}\sin\theta$$

$$F_y' = m\frac{d^2y}{dt^2}\cos\theta - m\frac{d^2x}{dt^2}\sin\theta \qquad (4.55)$$

and
$$F_z' = m\frac{d^2z}{dt^2}$$

The right side of Eq. (4.54) is evaluated by multiplying Eq. (4.51) by m and differentiating it twice with respect to time (assuming θ to be constant). This gives

$$m \frac{d^2 x'}{dt^2} = m \frac{d^2 x}{dt^2} \cos\theta + m \frac{d^2 y}{dt^2} \sin\theta$$

$$m \frac{d^2 y'}{dt^2} = m \frac{d^2 y}{dt^2} \cos\theta - m \frac{d^2 x}{dt^2} \sin\theta \qquad (4.56)$$

and $\qquad m \dfrac{d^2 z'}{dt^2} = m \dfrac{d^2 z}{dt^2}$

The right hand sides of Eqs. (4.55) and (4 56) are identical showing that if Newton's laws are true in A's frame of reference, they are also valid in B's frame which is rotated through a fixed angle with respect to the A's frame. This proves the invariance (symmetry) of Newton's laws for rotation of axes. This result has certain important consequences. First, no one can claim that his particular axes are unique, however, they may be more convenient for certain particular problems. Secondly that any piece of equipment which is completely self-contained with all the force-generating equipment completely inside the apparatus, would work the same way when turned at an angle. This is because the laws of physics governing the equipment remain the same. In chapter 1 (section 1.14) we have already proved the symmetry in linear motion. Therefore, Newton's laws are symmetrica lwith respect to linear and rotatory motion. This result is not true for Newton's laws alone. The other physical laws also exhibit the property of invariance (symmmetry) under translation and rotation of axes. These properties are so important that a mathematical technique has been developed on their basis for writing and using physical laws.

QUESTIONS AND PROBLEMS

4.1. Define torque and angular momentum. Prove that the torque is equal to the rate of change of angular momentum.

4.2.(a) Define moment of inertia of a rotating body. What is its physical significance?

(b) Prove the following relations:

(i) $J = I\omega$

(ii) $\tau = I\alpha$

where τ is the torque acting on a rigid body about a fixed axis and J, I, ω and α are, respectively, its angular momentum, moment of inertia, angular speed and angular acceleration.

4.3. State and prove the following theorems of moment of inertia:

(i) Theorem of parallel axes.

(ii) Theorem of perpendicular axes.

4.4. Calculate the moment of inertia of a solid uniform bar of rectangular cross-section about an axis passing through its centre of gravity and perpendicular to its length.

4.5. Determine the moment of inertia of a solid cylinder (i) about its axis of symmetry and (ii) about an axis normal to its axis of symmetry and passing through its centre of mass.

4.6. Derive the expression for the moment of inertia of a solid sphere (i) about a diameter and (ii) about a tangent.

4.7. State and prove the law of conservation of angular momentum. Discuss some important examples of this conservation principle.

4.8. Write an essay on rotational symmetry in physics.

4.9. A flywheel of mass 500 kg and 2 metres diameter makes 500 revolutions per minute. Assuming the mass to be concentrated at the rim, calculate the angular velocity, the moment of inertia and the energy of the flywheel.

(P.U. 1967, Delhi Univ.)

($50\pi/3$ rad/sec, 5×10^9 gm-cm², 6.861×10^{12} ergs)

[Hint: As the mass is concentrated at the rim, the radius of gyration of the flywheel is equal to its radius.]

4.10. Show that the rise of temperature of the earth, if it suddenly stops rotating, is given by

$$t = \tfrac{1}{5} \frac{R^2\omega^2}{JS}$$

where R, ω and S are the radius, angular speed, and specific heat of the earth respectively, and J is the mechanical equivalent of heat.

[Hint: Assume the earth to be a solid sphere.]

4.11. What will be the change in the duration of the day if the earth contracts suddenly to half its present radius?

(18 hrs)

4.12. The maximum and the minimum distances of a comet from the sun are 1.4×10^{12} m and 7×10^{10} m. If its velocity nearest to the sun is 6×10^4 m/sec, what is its velocity when farthest? Assume, in both positions, that the comet is moving in a circular orbit. [Agra 1970 S. D.U. 1980]

(3000 m/sec)

5. Gravitation

5.1. INTRODUCTION

Until the seventeenth century the tendency of a body to fall towards the earth was regarded as an inherent property of all bodies needing no further explanation. It was Issac Newton, a 23-year old scholar of Cambridge, who attempted to seek an explanation of the property of a body to fall towards the earth. He was inspired when he saw an apple falling from a tree while sitting in his garden at Woolsthrope. He immediately put a question to himself. Why should an apple or a body fall towards the earth? Whether the force, which attracted an apple to the earth, might also attract the moon to the earth? In his quest to answer these questions, Newton was the first person to realise that all bodies are attracted by the earth and that the attraction of the earth for bodies was not a peculiar property of the earth alone, but it was common to all bodies in the universe. He stated that every body attracted every other body irrespective of its shape or size. The bodies may be two balls lying on a table or two stars separated by thousands of miles.

The law governing the attraction between various bodies was first enunciated by Newton and is known as Newton's law of gravitation. It states:

"Every body in this universe attracts every another body with a force which is directly proportional to the product of their masses and inversely proportional to the square of the distance between them, the line of action of the force being along the line joining the centres of the two bodies."

Mathematically,

$$F = G\frac{m_1 m_2}{r^2} \qquad (5.1)$$

where F is the force of attraction (gravitational force) between two bodies of masses m_1 and m_2 separated by a distance r. G is a universal constant called the constant of gravitation.

If $m_1 = m_2 = 1$ and $r = 1$, Eq. (5.1) gives

$$F = G$$

Hence the gravitational constant is numerically equal to the force of attraction between two unit masses separated by a unit distance. Value of G is 6.669×10^{-8} C.G.S. units. As G is small, the force of attraction is not perceptible for small bodies but it becomes quite appreciable for large bodies like the sun, the stars and the planets. Since the force of attraction varies inversely as the square of the distance, the law of gravitation is also known as inverse square law.

Newton's law of gravitation can also be expressed in the vector form; that is,

$$\overrightarrow{F} = - G\frac{m_1 m_2}{r^3} \overrightarrow{r} \qquad (5.1a)$$

where $\overrightarrow{r}$ is the displacement vector between m_1 and m_2. Negative sign indicates that $\overrightarrow{F}$ and $\overrightarrow{r}$ are oppositely directed.

5.2. GRAVITATIONAL FIELD AND GRAVITATIONAL POTENTIAL

The space around a body within which its gravitational force can be experienced is called its gravitational field. The force experienced by a unit mass in a gravitational field is called the gravitational intensity at that point. In Eq. (5.1), if we take $m_1 = 1$ and $m_2 = M$, then the gravitational intensity is given by

$$I = \frac{GM}{r^2} \qquad (5.2)$$

The gravitational potential at a point in a gravitational field is defined as the amount of work done in carrying a unit mass from infinity up to that point against the gravitational intensity. In other words, the gravitational potential at a point is given by the potential energy per unit mass. Gravitational potential at a point distant R from a mass M is given by

$$V = \int\limits_{\infty}^{R} \frac{GM}{r^2}\, dr$$

or $$V = -\frac{GM}{R} \qquad (5.3)$$

Gravitational potential V and the gravitational intensity I are related as

$$I = -\frac{dV}{dx} \qquad (5.4)$$

where $\dfrac{dV}{dx}$ is the space rate of variation of potential.

Note that the potential V is a scalar while the intensity I is a vector quantity.

5.3. GRAVITATIONAL POTENTIAL AND FIELD DUE TO A THIN SPHERICAL SHELL

Consider a uniformly dense spherical shell of radius R and with centre at O. We shall derive the expressions for the gravitational potential and gravitational field due to the shell at a point P for the following cases:

1. *Point P Situated Outside the Shell.* Consider a point P situated at a distance r from the centre of a spherical shell whose potential and field are to be determined at P (Fig. 5.1). CF and

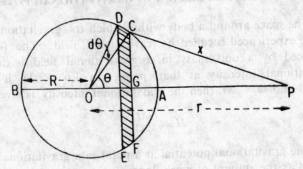

Fig. 5.1.   A thin spherical shell. Point P situated outside.

DE are two planes drawn normal to OP and lying very close to each other. A section of the shell cut out by these planes is as shown in Fig. 5.2.

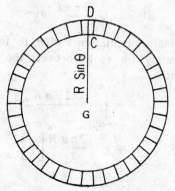

Fig. 5.2. Section of the shell cut out by planes DE and CR. $F\sin\theta$ is the radius of the ring (section).

If $\angle AOC = \theta$ and $\angle COD = d\theta$, the radius of the ring $CG = OC\sin\theta = R\sin\theta$. The ring width $DC = Rd\theta$. Therefore, the area of the ring = circumference × width

$$= 2\pi R\sin\theta \times Rd\theta$$
$$= 2\pi R^2 \sin\theta\, d\theta$$

If σ is the mass per unit area of the shell, the mass of the ring $= 2\pi R^2 \sigma \sin\theta\, d\theta$.

Using Eq. (5.3) the potential at P due to the elementary ring (whose every point is equidistant from P) is written as

$$dV = -\frac{G\,2\pi R^2 \sigma \sin\theta\, d\theta}{x} \tag{5.5}$$

where x is the distance of the ring from P.

In $\triangle OPC$

$$PC^2 = OC^2 + OP^2 - 2OC.OP\cos\theta$$

or $\qquad x^2 = R^2 + r^2 - 2Rr\cos\theta \tag{5.6}$

Differentiating Eq. (5.6), we get

$$2x\,dx = 2Rr\sin\theta\, d\theta$$

or $\qquad \sin\theta\, d\theta = \dfrac{x\,dx}{Rr} \tag{5.7}$

Substituting Eq. (5.7) into Eq. (5.5), we write

$$dV = -G\,\frac{2\pi R^2 \sigma\, x\,dx}{x \times Rr}$$

or $$dV = -G\frac{2\pi\sigma R}{r}dx \qquad (5.8)$$

The gravitational potential at P due to the entire shell is obtained by integrating Eq. (5.8) between the limits $x = r - R$ and $x = r + R$; that is,

$$V = \int_{r-R}^{r+R} - G\frac{2\pi\sigma R}{r}dx$$

$$= -\frac{G2\pi\sigma R}{r}\Big[x\Big]_{r-R}^{r+R}$$

$$= -\frac{G2\pi\sigma R}{r}2R$$

or $$V = -\frac{G4\pi R^2\sigma}{r}$$

Hence $$V = -\frac{GM}{r} \qquad (5.9)$$

where $M\,(= 4\pi R^2\sigma)$ is the mass of the shell. Thus the gravitational potential at a point outside a spherical shell behaves as if the entire mass of the shell is concentrated at its centre.

The gravitational intensity at P due to the shell is given by

$$I = -\frac{dV}{dr} = -\frac{d}{r}\left(-\frac{GM}{r}\right) = -\frac{GM}{r^2} \qquad (5.10)$$

Hence the force acting on a point mass m at P is expressed as

$$F = mI = -\frac{GMm}{r^2} \qquad (5.11)$$

2. *Point P Situated on the Surface of the Shell.* The gravitational potential at a point situated on the surface of a spherical shell is obtained by integrating Eq. (5.8) between the limits $x = 0$ and $x = 2R$; that is

$$V = \int_{0}^{2R} - G\frac{2\pi R\sigma}{r}dx = -\frac{G2\pi R\sigma}{r}\Big[x\Big]_{0}^{2R}$$

$$= -G\frac{4\pi R^2\sigma}{R} \quad [\text{Here } r = R]$$

or $$V = -\frac{GM}{R} \qquad (5.12)$$

The gravitational intensity at a point on the surface of the shell is given by

$$I = -\frac{dV}{dr} = -\frac{GM}{R^2} \qquad (5.13)$$

Thus a uniformly dense spherical shell attracts a point mass on its surface as if its mass is concentrated at its centre.

Force acting on a point mass m at P is

$$F = mI = -\frac{GMm}{R^2} \qquad (5.14)$$

3. *Point P Situated Inside the Shell.* The gravitational potential at a point P situated inside a spherical shell (Fig. 5.3) is obtained

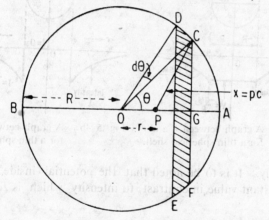

Fig. 5.3. A spherical shell. Point P situated inside the shell.

by integrating Eq. (5.8) between the limits $x = R - r$ and $x = R + r$; that is,

$$V = \int_{R-r}^{R+r} -\frac{G2\pi R\sigma}{r}\, dx = -G\frac{2\pi R\sigma}{r}\left[x \right]_{R-r}^{R+r}$$

$$= -\frac{G2\pi R\sigma}{r}2r = -\frac{G4\pi R^2\sigma}{R}$$

or

$$V = -\frac{GM}{R} \qquad (5.15)$$

Eq. (5.15) shows that the potential at any point inside a spherical shell is constant and is the same as the potential on its surface.

Gravitational intensity I at

at $$P = -\frac{dV}{dr} = -\frac{d}{dr}\left(-\frac{GM}{R}\right) = 0. \qquad (5\ 15a)$$

Gravitational force on a point mass m at

$$P = F = mI = 0 \qquad (5.16)$$

Eqs. (5.15a) and (5.16) show that the gravitational intensity and gravitational force at a point inside a spherical shell is always zero.

The graphs showing the variations of gravitational potential and gravitational intensity with the distance r of a point from the centre of a thin spherical shell are shown in Figs. 5.4a and 5.4b

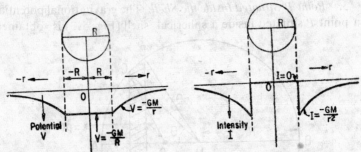

Fig. 5.4a.   A graph between V and r Fig. 5.4b.   A graph between I and r
for a thin spherical shell.   for a thin spherical shell.

respectively. It is to be noted that the potential inside the shell has a constant value in contrast to intensity which is zero inside the shell.

5.4.   GRAVITATIONAL POTENTIAL AND FIELD DUE TO A SOLID SPHERE

1. *Point P Situated Outside the Sphere.* Consider a point P at a distance r from the centre of a solid sphere of radius R and centre O (Fig. 5.5). We have to find the value of the gravitational potential at P. Divide the solid sphere into a large number of imaginary thin spherical shells of masses m_1, m_2, m_3 etc. (Fig. 5.5).

Using Eq. (5.9) the potential at P due to each individual shell will be given by $\dfrac{-Gm_1}{r}, \dfrac{-Gm_2}{r}, \dfrac{-Gm_3}{r}$ etc. Now the potential at P due to the solid sphere is equal to the sum of the potentials at P due to the elementary shells. Hence the gravitational potential at P due to the solid sphere is given by

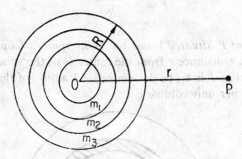

Fig. 5.5. A solid sphere of radius R and centre at O. Point P is situated outside the sphere.

$$V = -\frac{Gm_1}{r} - \frac{Gm_2}{r} - \frac{Gm_3}{r} \ldots$$

$$= -\frac{G}{r}(m_1 + m_2 + m_3 + \ldots)$$

or $\qquad V = -\dfrac{GM}{r}$ $\qquad\qquad$ (5.17)

where $M(= m_1 + m_2 + m_3 \ldots)$ is the mass of the sphere.

Thus the gravitational potential at a point outside a solid sphere acts as if the entire mass of the sphere is concentrated at its centre.

Gravitational intensity at $P = I = -\dfrac{dV}{dr} = -\dfrac{GM}{r^2}$ $\qquad$ (5.18)

Gravitational force on a point mass m at P is

$$F = mI = -\frac{GMm}{r^2} \qquad\qquad (5.19)$$

2. *Point P Situated on the Surface of the Solid Sphere.* Obviously the potential at a point P on the surface of the solid sphere will be given by

$$V = -\frac{GM}{r} = -\frac{GM}{R} \text{ (as } r = R \text{ on the surface)} \qquad (5.20)$$

and the gravitational intensity and force are

$$I = -\frac{GM}{R^2}$$

$$F = -\frac{GMm}{R^2} \tag{5.21}$$

3. *Point P Situated Inside the Solid Sphere.* Consider a point P situated at a distance r from the centre O of the sphere for which the potential at P is to be determined. Radius of the sphere is R and its mass per unit volume is ρ.

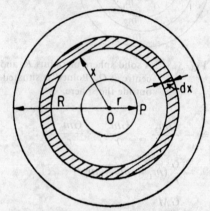

Fig. 5.6. A solid sphere of radius R and centre O. Point P is situated inside.

With radius r draw an imaginary sphere through P. This will divide the solid sphere into an inner solid sphere of radius r and an outer spherical shell of inner radius r and outer radius R. Using Eq. (5.20) the potential at P due to the inner solid sphere is given by

$$V_1 = -G\frac{\frac{4}{3}\pi r^3 \rho}{r}$$

or $\qquad\qquad\qquad V_1 = -\tfrac{4}{3}\pi r^2 \rho G \tag{5.22}$

In order to determine the potential at P due to the outer spherical shell, we imagine the outer shell to be divided into a number of thin concentric shells. Consider one such concentric shell of radius x and thickness dx (Fig. 5.6).
Now the volume of this elementary shell

$$= 4\pi x^2 \, dx$$

Therefore the mass of the elementary shell

$$= 4\pi x^2 \rho \, dx$$

The gravitational potential at P due to this elementary shell is given by Eq. (5.15); that is,

$$dV_2 = -G\frac{4\pi x^2 \rho \, dx}{x}$$

$$dV_2 = -4\pi \rho G x \, dx \qquad (5.23)$$

The potential at P due to the entire outer spherical shell is obtained by integrating Eq. (5.23) between the limits $x = r$ and $x = R$; that is,

$$V_2 = \int_r^R -4\pi \rho G \, x \, dx$$

$$= -4\pi \rho G \left[\frac{x^2}{2}\right]_r^R$$

or $\qquad V_2 = -2\pi \rho G (R^2 - r^2) \qquad (5.24)$

Therefore the potential at a point P inside a solid sphere is given by the sum of Eqs. (5.22) and (5.24); that is,

$$V = V_1 + V_2 = -\tfrac{4}{3}\pi r^2 G\rho - 2\pi \rho G(R^2 - r^2)$$

$$= -2\pi \rho G\left(\frac{2r^2}{3} + R^2 - r^2\right)$$

$$= -\tfrac{2}{3}\pi \rho G(3R^2 - r^2)$$

$$= -\tfrac{4}{3}\pi R^3 \rho G\,\frac{(3R^2 - r^2)}{2R^3}$$

or $\qquad V = -GM\dfrac{(3R^2 - r^2)}{2R^3} \qquad (5.25)$

where $M(= \tfrac{4}{3}\pi R^3 \rho)$ is the mass of the sphere.

The gravitational intensity at the point P is given by

$$I = -\frac{dV}{dr} = -\frac{d}{dr}\left[\frac{-GM(3R^2 - r^2)}{2R^3}\right]$$

or $\qquad I = -\dfrac{GM}{R^3}r \qquad (5.26)$

Thus the gravitational intensity at a point inside the solid sphere varies directly as the distance of the point from the centre of the sphere. Intensity is zero at the centre of the sphere.

Force acting on a point mass m at P is

$$F = mI = -\frac{GMm}{R^3}r \qquad (5.27)$$

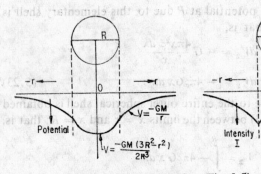

Fig. 5.7a. A graph between V and r for a solid sphere.

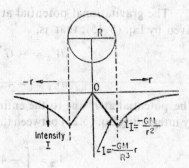

Fig. 5.7b. A graph between I and r for a solid sphere.

The graphs between potential and distance and between intensity and distance for the case of a solid sphere are shown in Figs. 5.7a and 5.7b respectively. Fig. 5.7a shows that the magnitude of the potential continuously increases with decrease in r and is maximum $\left(=-\frac{3}{2}\frac{GM}{R}\right)$ at the centre of the sphere. Fig. 5.7b tells us that the intensity increases with decrease in the value of r up to the surface of the sphere where it becomes maximum. Within the sphere, the intensity decreases linearly till it becomes zero at the centre of the sphere.

5.5. ORBITAL AND ESCAPE VELOCITIES

Consider a body P moving round the earth in a circular orbit of radius r (Fig. 5.8). The gravitational attraction of the earth on P provides the necessary centripetal force to keep the body in motion round E.

Now centripetal force $= \dfrac{mV_0^2}{r}$ and gravitational force $= \dfrac{GMm}{r^2}$

where V_0 is the orbital velocity, m and M are the masses of the body P and the earth respectively, and G is the constant of gravitation. Therefore

$$\frac{mV_0^2}{r} = \frac{GMm}{r^2}$$

or

$$V_0^2 = \frac{GM}{r}$$

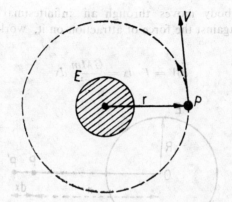

Fig. 5.8. A body P moving round the earth
E in a circular orbit of radius r.

Hence $$V_0 = \sqrt{\frac{GM}{r}} \qquad (5.28)$$

Eq. (5.28) gives the orbital velocity of a body at a distance r
from the centre of the earth. If the body is moving near the surface of the earth, $r \approx R$, the radius of the earth. Thus the orbital
velocity of a body near the surface of the earth is given by

$$V_0 = \sqrt{\frac{GM}{R}} \qquad (5.29)$$

When a body is projected upwards, it reaches a particular
height depending on the velocity of projection and then it comes
back to the earth due to the gravitational attraction. But if the
velocity of projection of the body is such that the kinetic energy
possessed by the body exceeds the gravitational pull of the earth,
then it will never return to the earth and will escape into the
atmosphere. The minimum velocity, with which a body should
be projected in order that it may escape into the atmosphere, is
called the escape velocity of the body.

Consider a body of mass m situated at a distance x from
earth's centre ($x > R$, the radius of the earth)

From Newton's law of gravitation

$$F = G\frac{Mm}{x^2}$$

where M is the mass of the earth and F is the force of attraction
between M and m.

If the body moves through an infinitesimal distance dx (Fig. 5.9a) against the force of attraction on it, work done by it is expressed as

$$dW = F \cdot dx = \frac{GMm}{x^2} \, dx \qquad (5.30)$$

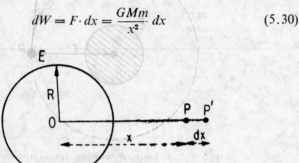

Fig. 5.9a. A body at a distance x from the centre of the earth. $PP' = dx$.

Hence work done to move the body from earth's surface to infinity against the gravitational attraction is obtained by integrating Eq. (5.30) between the limits $x = R$ and $x = \infty$. Thus

$$W = \int_{R}^{\infty} \frac{GMm}{x^2} \, dx = GMm \left[-\frac{1}{x} \right]_{R}^{\infty}$$

or
$$W = \frac{GMm}{R} \qquad (5.31)$$

In order that the body may escape from the earth, its kinetic energy must be equal to Eq. (5.31); that is,

$$\frac{1}{2} mV_e^2 = \frac{GMm}{R}$$

or
$$V_e = \sqrt{\frac{2GM}{R}} \qquad (5.32)$$

where V_e is the escape velocity of the body. Note that the escape velocity for all the bodies from the earth is constant. Substituting the values of G, M and R in Eq. (5.32), the escape velocity of a body is found to be 11.2 km/sec.

5.6. MASS AND MEAN DENSITY OF THE EARTH

Let a body of mass m be situated on the surface of the earth.

The force with which this body is attracted towards the earth is given by Newton's law of gravitation and is equal to the weight of the body; that is,

$$F = mg = G\frac{Mm}{R^2} \qquad (5.33)$$

where g is the acceleration due to gravity on earth's surface, M and R are the mass and the radius of the earth respectively.

Eq. (5.33) gives

$$g = \frac{GM}{R^2} \qquad (5.34)$$

and

$$M = \frac{gR^2}{G} \qquad (5.35)$$

Using Eq. (5.35) mass of the earth can be calculated.

Also, the mass of the earth is given by

$$M = \text{Volume of the earth} \times \text{Mean density of the earth}$$

or

$$M = \tfrac{4}{3}\pi R^3 \times \rho \qquad (5.36)$$

where ρ is the mean density of the earth. Substituting Eq. (5.36) in Eq. (5.35), the mean density of the earth is written as

$$\rho = \frac{3g}{4\pi RG} \qquad (5.37)$$

Let us calculate the value of acceleration due to gravity at a point situated inside the surface of the earth. Assume the earth to be a solid sphere of mass M, radius R and mean density ρ. Consider a mass m lying at a depth h inside the earth. Let g and g' be the accelerations due to gravity on the earth's surface and inside the earth's surface respectively.

Using Eq. (5.27), we have

$$F = -mg' = -\frac{GMm(R-h)}{R^3} \qquad (5.38)$$

or

$$g' = \frac{GM(R-h)}{R^3}$$

$$= \frac{G\tfrac{4}{3}\pi R^3 \rho (R-h)}{R^3}$$

Thus,

$$g' = \tfrac{4}{3}\pi G\rho (R-h) \qquad (5.39)$$

Using Eq. (5.34) in Eq. (5.38), we write

$$g' = \frac{g(R - h)}{R} \qquad (5.40)$$

The negative sign in Eq. (5.38) signifies that the force of attraction is directed towards the centre of the earth.

5.7. KEPLER'S LAWS OF PLANETARY MOTION

Many scientific attempts have been made to understand and explain the movements of celestial bodies. In the second century Ptolemy proposed a theory known as Ptolemaic or *geocentric theory*. This theory assumed the earth to be stationary at the centre of the universe and the other celestial bodies revolving round it in complete orbits. In the sixteenth century Copernicus suggested another theory known as the heliocentric theory. According to this theory the sun was supposed to be at rest and situated at the centre of the universe. The other celestial bodies were rotating about the sun in various orbits. Growing controversy over these two conflicting theories stimulated and led Tycho Brahe and his assistant Kepler to make more accurate observations to explain the celestial motions. After a long analysis and deep interpretation of his master's results on the motion of celestial bodies, Kepler formulated three empirical laws known as Kepler's laws of planetary motion.

1. *Kepler's First Law*. This law is also known as the law of *elliptical orbits*. *It states that all planets move in elliptical orbits having the sun at one focus.*

2. *Kepler's Second Law*. This is also known as the law of *equal areas*. *It states that the line joining any planet to the sun sweeps out equal areas in equal interval of time*; that is, the areal velocity of the radius vector is constant.

3. *Kepler's Third Law*. This is also known as the law of *periods* or the *harmonic law*. *It states that the square of the time period of any planet about the sun is proportional to the cube of the planet's mean distance from the sun.*

Helio-centric theory of Copernicus was strongly supported by Kepler's laws which showed that planetary motions could be described with great simplicity when the sun was taken as the reference body. However, these laws were empirical in character as they simply described the observed motion of the planets without any theoretical proof. At that time the concept of force had not been clearly formulated and it was left to the genius of Sir Issac Newton to give a theoretical interpretation of Kepler's laws through his laws of motion and his law of gravitation. In this way Newton was able to explain the planetary motions in the solar system and also terrestrial motions with one common concept, thus synthesizing into one theory the previously separate sciences of terrestrial mechanics and celestial mechanics.

4. *Proof of Kepler's Second Law.* Figure 5.9b shows a planet *P* revolving about the sun *S* along some arbitrary path. The area swept out by the radius vector in a very short time interval Δt is

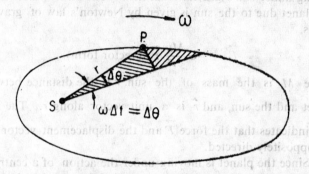

Fig. 5.9b. A planet moving along an elliptical path with the sun *S* at the focus of the ellipse. In time Δt the planet sweeps out an angle $\Delta\theta = \omega\,\Delta t$.

shown shaded in the figure. This area, neglecting the small triangular region at the end, is one-half the base times the altitude; that is,

$$\text{the area swept out} \approx \tfrac{1}{2} r^2\,\Delta\theta$$

Hence the areal velocity of the radius vector is expressed as

$$\operatorname*{Lt}_{\Delta t \to 0} \tfrac{1}{2} r^2 \frac{\Delta\theta}{\Delta t} = \tfrac{1}{2} r^2 \frac{d\theta}{dt}$$

or $$\text{areal velocity} = \tfrac{1}{2} r^2 \omega \qquad\qquad (5.41)$$

where $\omega\,(= d\theta/dt)$ is the angular velocity of *P* round the sun.

Now the planet *P* is moving around the sun under the action of a central force (gravitational attraction between *P* and *S*).

Hence the angular momentum J of the planet P is conserved; that is,

$$J = mr^2\omega = \text{constant}$$

where m is the mass of the planet P. Since mass m of the planet is a constant quantity, we have

$$r^2\omega = h = \text{constant} \tag{5.42}$$

Using Eq. (5.42) into Eq. (5.41), we get

$$\text{areal velocity} = \text{constant}$$

Proof of Kepler's First Law. Consider a planet of mass m moving under sun's gravitational field. The central force acting on the planet due to the sun is given by Newton's law of gravitation; that is,

$$\vec{F} = - G\frac{Mm}{r^2}\hat{r} \quad \text{(vector form)} \tag{5.43}$$

where M is the mass of the sun, r is the distance between the planet and the sun, and $\hat{r}$ is a unit vector along r. The negative sign indicates that the force $\vec{F}$ and the displacement vector $\vec{r}\ (= r\hat{r})$ are oppositely directed.

Since the planet is moving under the action of a central force, it possesses only the radial component of the acceleration. Hence applying Newton's second law of motion, the force acting on the planet is expressed as

$$\vec{F} = m\vec{a}_R = m\left[\frac{d^2r}{dt^2} - r\left(\frac{d\theta}{dt}\right)^2\right]\hat{r} \tag{5.44}$$

where $\vec{a}_R\left[= \left\{\frac{d^2r}{dt^2} - r\left(\frac{d\theta}{dt}\right)^2\right\}\hat{r}\right]$ is the radial component of the acceleration.

From Eqs. (5.43) and (5.44), we get

$$m\left[\frac{d^2r}{dt^2} - r\left(\frac{d\theta}{dt}\right)^2\right]\hat{r} = - G\frac{Mm}{r^2}\hat{r}$$

or

$$\frac{d^2r}{dt^2} - r\left(\frac{d\theta}{dt}\right)^2 = -\frac{GM}{r^2}$$

or

$$\frac{d^2r}{dt^2} - r\omega^2 = -\frac{GM}{r^2}$$

Multiplying both sides by r^3, we write

$$r^3\frac{d^2r}{dt^2} - r^4\omega^2 = -GMr \qquad (5.45)$$

Using Eq. (5.42) into Eq. (5.45), we have

$$r^3\frac{d^2r}{dt^2} - h^2 = -GMr \qquad (5.46)$$

Let $r = \frac{1}{u}$, then

$$\frac{dr}{dt} = \frac{d}{dt}\left(\frac{1}{u}\right) = -\frac{1}{u^2}\frac{du}{dt} = -\frac{1}{u^2}\frac{du}{d\theta}\cdot\frac{d\theta}{dt}$$

or $$\frac{dr}{dt} = -h\frac{du}{d\theta} \qquad \left(\because \frac{1}{u^2}\frac{d\theta}{dt} = r^2\omega = h\right)$$

Therefore, $$\frac{d^2r}{dt^2} = -h\frac{d}{dt}\left(\frac{du}{d\theta}\right) = -h\frac{d}{d\theta}\left(\frac{du}{d\theta}\right)\frac{d\theta}{dt}$$

or $$\frac{d^2r}{dt^2} = -h\frac{d^2u}{d\theta^2}\cdot\frac{d\theta}{dt} = -h\omega\frac{d^2u}{d\theta^2}$$

or $$\frac{d^2r}{dt^2} = -h^2u^2\frac{d^2u}{d\theta^2} \qquad (5.47)$$

Substituting Eq. (5.47) in Eq. (5.46), we get

$$\frac{1}{u^3}\left(-h^2u^2\frac{d^2u}{d\theta^2}\right) - h^2 = -\frac{GM}{u}$$

or $$\frac{d^2u}{d\theta^2} + u = \frac{GM}{h^2}$$

or $$\frac{d^2}{d\theta^2}\left(u - \frac{GM}{h^2}\right) + \left(u - \frac{GM}{h^2}\right) = 0 \qquad (5.48)$$

$$\left(\because \frac{GM}{h^2} = \text{constant}\right)$$

Eq. (5.48) is a differential equation of the standard form

$$\frac{d^2y}{d\theta^2} + y = 0$$

and, therefore, its solution is given by

$$u - \frac{GM}{h^2} = A\cos\theta \qquad (5.49)$$

where A is a constant which can be determined by boundary conditions.

Eq. (5.49) gives

$$\frac{1}{r} - \frac{GM}{h^2} = A \cos \theta$$

or

$$\frac{1}{r} = \frac{GM}{h^2} + A \cos \theta$$

or

$$\frac{h^2/GM}{r} = 1 + \frac{Ah^2}{GM} \cos \theta$$

or

$$\frac{l}{r} = 1 + e \cos \theta \qquad (5.50)$$

Eq. (5.50) is the equation of an ellipse whose semi latus rectum is $l \ (= h^2/GM)$ and eccentricity is $e \ (= Ah^2/GM < 1)$.

Hence the path of the planet round the sun is an ellipse. So Kepler's first law is established.

Proof of Kepler's Third Law. If T is the time period of a planet round the sun, then

$$T = \frac{\text{area of the ellipse}}{\text{areal velocity}}$$

or

$$T = \frac{\pi ab}{\frac{1}{2}r^2\omega}$$

where a and b are the semi-major and semi-minor axes of the ellipse respectively.

Therefore,

$$T = \frac{2\pi ab}{h}$$

or

$$T^2 = \frac{4\pi^2 a^2 b^2}{h^2} \qquad (5.51)$$

Now the semi latus rectum of the ellipse is given by

$$l = \frac{b^2}{a} = \frac{h^2}{GM}$$

Hence

$$b^2 = \frac{h^2 a}{GM}$$

Substituting the value of b^2 in Eq. (5.51), we get

$$T^2 = \frac{4\pi^2 a^2 h^2 a}{h^2 GM}$$

or
$$T^2 = \frac{4\pi^2}{GM} a^3$$

or
$$T^2 = Ka^3 \qquad (5.52)$$

where $K \, (= 4\pi^2/GM)$ is a constant.

Hence
$$T^2 \propto a^3$$

This establishes Kepler's third law of planetary motion.

5.8. DERIVATION OF NEWTON'S LAW OF GRAVITATION FROM KEPLER'S LAWS

According to Kepler's first law, the path of a planet round the sun is an ellipse whose equation is given by Eq. (5.50); that is,

$$\frac{l}{r} = 1 + e \cos\theta$$

or
$$lu = 1 + e \cos\theta \left[\because u = \frac{1}{r} \right] \qquad (5.53)$$

Differentiating Eq. (5.53) twice with respect to time, we have

$$l\frac{d^2u}{d\theta^2} = - e \cos\theta \qquad (5.54)$$

Adding Eqs. (5.53) and (5.54), we write

$$l\left(u + \frac{d^2u}{d\theta^2} \right) = 1$$

or
$$u + \frac{d^2u}{d\theta^2} = \frac{1}{l} \qquad (5.55)$$

A planet moving round the sun can have two components of acceleration. They are the radial and the tangential components, and are, respectively, given by

$$\vec{a}_R = \left[\frac{d^2r}{dt^2} - r\left(\frac{d\theta}{dt}\right)^2 \right] \hat{r}$$

and
$$\vec{a}_T = \left[\frac{1}{r}\frac{d}{dt}\left(r^2 \frac{d\theta}{dt} \right) \right] \hat{p}$$

where $\hat{r}$ and $\hat{p}$ are unit vectors along and perpendicular to r respectively.

According to Kepler's second law, the areal velocity

$$\left(= \frac{1}{2} r^2 \frac{d\theta}{dt} \right)$$

is constant and hence the tangential component of the acceleration acting on the planet is zero; that is, $\vec{a_T} = 0$. Therefore, the planet moves round the sun under the action of only the radial component of acceleration; that is,

$$\vec{a_R} = \left[\frac{d^2r}{dt^2} - r\left(\frac{d\theta}{dt}\right)^2\right]\hat{r}$$

Using Eq. (5.47) in the above expression, we get

$$\vec{a_R} = \left[-h^2u^2\frac{d^2u}{d\theta^2} - \frac{1}{u}(hu^2)^2\right]\hat{r} \qquad \left[\because \frac{d\theta}{dt} = \omega = hu^2\right]$$

or $\qquad \vec{a_R} = -h^2u^2\left[u + \frac{d^2u}{d\theta^2}\right]\hat{r}$ (5.56)

Substituting Eq. (5.55) in Eq. (5.56), the expression for the radial acceleration becomes

$$\vec{a_R} = -\frac{h^2u^2}{l}\hat{r}$$

$$= -\frac{h^2}{l}\cdot\frac{1}{r^2}\hat{r}$$

If we put $\frac{h^2}{l} = K'$, then

$$\vec{a_R} = -\frac{K'}{r^2}\hat{r}$$ (5.57)

Thus the radial acceleration varies inversely as the square of the distance between the planet and the sun; the negative sign signifies that $\vec{a_R}$ and $\vec{r}$ are oppositely directed; that is, the acceleration is directed towards the sun.

Now the force acting on the planet due to the sun is given by

$$\vec{F} = m\vec{a_R} = -\frac{K'm}{r^2}\hat{r}$$ (5.58)

According to **Kepler's** third law

$$T^2 \propto a^3$$

or $\qquad\qquad T^2 = Ka^3$ (5.59)

where K is a constant of proportionality and is constant for all the planets.

Also

$$T^2 = \frac{4\pi^2 a^2 b^2}{h^2} \quad \text{[see eq. (5.51)]}$$

or

$$T^2 = \frac{4\pi^2 l}{h^2} a^3 \left(\because l = \frac{b^2}{a} \right)$$

Thus,
$$T^2 = \frac{4\pi^2}{K'} a^3 \tag{5.60}$$

Comparing Eqs (5.59) and (5.60), we get

$$K = \frac{4\pi^2}{K'}.$$

This means that K' also has the same value for all the planets.

Since the force of attraction $\vec{F}$ between the planet and the sun is mutual, this force of attraction will also be proportional to the mass of the sun; that is,

$$\vec{F} \propto M \tag{5.61}$$

where M is the mass of the sun. Combining Eq. (5.61) with Eq. (5.58), we can write

$$\vec{F} \propto -\frac{Mm}{r^2} \hat{r}$$

or
$$\vec{F} = -G \frac{Mm}{r^2} \hat{r}$$

which is nothing but Newton's law of gravitation. Here $G \left(= \frac{K'}{M} \right)$

is the constant of proportionality called the constant of gravitation. Note that $K'(= GM)$ is constant for all the planets.

The deduction of Newton's law of gravitation given above is only approximate as the sun is not stationary but is moving about the centre of mass of the sun and the planet. Actually the effective mass of the planet will be $\frac{mM}{M + m}$ and not m. But as the sun is too heavy ($M \gg m$) the effect of its motion can be neglected.

5.9. DETERMINATION OF THE CONSTANT OF GRAVITATION

Methods for the determination of the constant of gravitation G can be classified into two groups:

(1) *Large Scale Methods.* Mountain experiment and Airy's Mine method are the large-scale methods used to determine the value of G.

(2) *Laboratory Methods.* They consist of Cavendish method, Boy's method and Poynting's method (weighing method).

Large scale methods depend on the measurement of the force of attraction exerted by huge bodies like mountains and earth. As the measurement involved in these experiments are quite uncertain, they are only of historical importance.

The force of attraction between bodies of moderate sizes is very small. So we need a very sensitive instrument to measure it. This instrument known as "torsion balance" was invented by Rev. John Michell. Before Michell could use his invention to measure gravitational attraction, he died. The torsion balance of Michell with slight improvement was first used by Sir Henry Cavendish in 1798 to determine the value of G. The apparatus used by Cavendish is schematically indicated in Fig. 5.10. It consists of two smaller balls of lead (m, m) mounted at the opposite ends of a light horizontal rod RS which is supported at its centre by a fine torsion fibre. Two larger lead spheres (M, M) are brought up to the position shown. The forces of gravitational attraction between the large and the small spheres result in a couple which twists the fibre through a small angle which can be measured by a vernier and scale arrangement (not shown in Fig. 5.10). The system comes

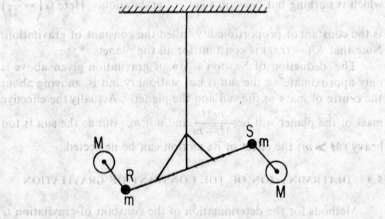

Fig. 5.10. Principle of Cavendish balance.

into equilibrium when the restoring couple due to twist in the
suspension wire balances the deflecting couple due to the force of
attraction between large and small spheres.

The deflecting couple $= G\dfrac{Mm}{d^2} \cdot 2l$ $\hspace{3cm}$ (5.62)

where d is the distance between the centres of a large ball and the
near small ball, $2l$ is the length of the rod carrying m, m, and G is
the constant of gravitation.

Restoring couple $= C\theta$. $\hspace{3.5cm}$ (5.63)

Here C is the restoring couple per unit twist of the suspension
wire and θ is the mean angular deflection of the rod RS.

In equilibrium

$$G\frac{Mm}{d^2} 2l = C\theta$$

or $\hspace{3cm}$ $G = \dfrac{Cd^2}{2Mml}\theta$ $\hspace{3cm}$ (5.64)

C is determined by calculating the moment of inertia and the
time period of the vibrating system; that is,

$$T = 2\pi \sqrt{\frac{I}{C}}$$

Hence $\hspace{3cm}$ $C = \dfrac{4\pi^2 I}{T^2}$

Substituting the value of C in Eq. (5.64), we get

$$G = \frac{2\pi^2 Id^2}{MlmT^2}\theta \hspace{3cm} (5.65)$$

The value of G determined by Cavendish was found to be
6.562×10^{-8} C.G.S. units.

Cavendish experiment suffered from following weak points:

1. The apparatus was kept in a large chamber where the
temperature control was difficult. This resulted in convection
currents which produced erratic measurements of the system
causing an error in the measurement of deflection θ.

2. The suspension wire used to support the torsion rod was
such that the deflection produced was small.

3. The angle θ could not be measured accurately with the help of vernier and scale arrangement.

4. The rod supporting the larger spheres and the torsion rod also exerted attractive forces mutually and tended to affect the gravitational couple.

5. The suspension fibre used was not perfectly elastic and so the couple required to twist it was not proportional to deflection.

6. The attraction between the larger sphere and the distant smaller sphere was not taken into account.

Sir Charles V. Boys modified the apparatus so as to eliminate all the weak points which were present in Cavendish experiment. His method has been described below.

Boys' Method. The apparatus used by Boys to determine the value of G is as shown in Fig. 5.11. It consists of two coaxial tubes T_1 and T_2. T_1 is fixed while T_2 is capable of rotating about

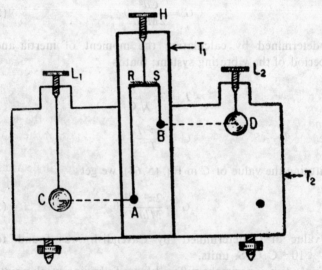

Fig. 5.11.   Boys' apparatus.

a common vertical axis. The inner tube T_1 contains a torsion rod in the form of a small mirror strip RS (2.5 cm long) attached to a torsion head H by means of a suspension wire made of a fine fused quartz fibre of length 43.3 cm and diameter 0.0125 mm. Two small gold spheres A and B each of about 0.5 cm diameter and

2.65 gm mass are suspended at the ends of the torsion rod by means of quartz fibres of unequal lengths. The vertical distance between the centres of A and B is nearly 15 cm. The advantage of using a quartz fibre is that it is almost perfectly elastic, is stronger than a steel wire of the same diameter and can be drawn very fine. Restoring torque is also very small.

Two attracting lead spheres C and D, each of mass 1407 gm and diameter 10.8 cm, are suspended respectively from the revolving lids L_1, L_2 of the tube T_2 such that the centres of A and C, and also those of B and D, are at the same level. The distance between the centres of A and C is the same as that between the centres of B and D. The deflection of the torsion rod is measured by lamp and scale arrangement; the strip RS serving as the reflector of light. The entire apparatus is quite compact and carries levelling screws as shown in Fig. 5.11.

The apparatus is first levelled with the help of the levelling screws. The revolving lids L_1 and L_2 are rotated so that C, D, A and B are in line with the strip RS. In this position no deflecting couple will act on the strip RS. Note the readings on the scale. Adjust C and D so that they are on the opposite sides of A and B respectively but not in line with RS. In this position the torsion rod RS will experience a deflecting couple due to the gravitational attraction between the large and small spheres although their centres are situated in different horizontal planes. Stop rotating C and D when maximum deflection is obtained on the scale. Let the deflection be θ_1. Next bring C and D on the other sides of A and B respectively and adjust the revolving lids by the same amount as in the first case so that the maximum deflection is obtained in the opposite direction. Let it be θ_2. Then the mean deflection of the torsion rod will be

$$\frac{\theta_1 + \theta_2}{2} = \theta \text{ (say)}$$

Fig. 5.12 shows the position of A, B, C and D in the equilibrium position of maximum deflection θ as viewed from the top of the apparatus. O is the mid point of strip RS whose length is $2l$. The distance between the centres of lead spheres C and D is $2a$. x is the distance between the centres of A and C or B and D.

Now the deflecting couple

$$= G \frac{Mm}{x^2} . EF$$

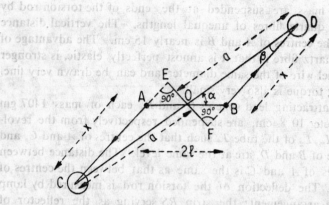

Fig. 5.12. Position of large and small spheres as viewed from the top of the apparatus. The system is in equilibrium after attaining a maximum deflection θ.

or   deflecting couple $= G\dfrac{Mm}{x^2}.2 0F$ (5.66)

where $EF(=2 0F)$ is the perpendicular distance between the lines AC and BD.

Let $\angle BOD = \alpha$ and $\angle BDO = \beta$.

In $\triangle\ BOD$,

$$BD^2 = x^2 = a^2 + l^2 - 2al \cos \alpha$$

or $x = [a^2 + l^2 - 2al \cos \alpha]^{1/2}$.   (5.67)

In right angled triangle DOF

$$OF = a \sin \beta$$

Also in triangle BOD,

$$\frac{x}{\sin \alpha} = \frac{l}{\sin \beta}$$

or $\sin \beta = \dfrac{l \sin \alpha}{x}$

Therefore,

$$OF = a \sin \beta = \frac{al \sin \alpha}{x}$$ (5.68)

Substituting Eqs. (5.67) and (5.68) in Eq. (5.66), we get

$$\text{deflecting couple} = \frac{2G\ Mm\ al \sin \alpha}{(a^2 + l^2 - 2al\cos \alpha)^{3/2}}$$ (5.69)

The deflecting couple will generate a restoring couple due to the twist in the suspension wire. The system will come into equilibrium when the deflecting couple is balanced by the restoring couple.

Now restoring couple $= C\theta$

where C is the restoring couple per unit deflection.

Hence in equilibrium,

$$\frac{2G\ Mm\ al\ \sin \alpha}{(a^2 + l^2 - 2al\ \cos \alpha)^{3/2}} = C\theta$$

or $$G = \frac{(a^2 + l^2 - 2al\ \cos \alpha)^{3/2} C\theta}{2Mm\ al\ \sin \alpha} \qquad (5.70)$$

Time period of the oscillating system (with mirror strip and gold spheres) in the absence of lead spheres is given by

$$T = 2\pi \sqrt{I/C}$$

where I is the moment of inertia of the moving system about its vertical axis. Thus,

$$T^2 = 4\pi^2 \frac{I}{C}$$

or $$C = \frac{4\pi^2 I}{T^2} \qquad (5.71)$$

Using Eq. (5.71) into Eq. (5.70), we write

$$G = \frac{2\pi^2 I (a^2 + l^2 - 2al\ \cos \alpha)^{3/2}}{Mm\ al\ T^2 \sin \alpha} \theta \qquad (5.72)$$

Eq. (5.72) is used to determine the value of G. The value of G as calculated by this method is 6.6576×10^{-8} C.G.S. units.

Boys' method has the following advantages over Cavendish method:

(1) Because of the small size of Boys' apparatus it is easy to control the temperature and thus eliminate convection currents.

(2) For quartz suspension the deflection is large and the restoring couple is proportional to deflection. Also, because of its high elasticity the quartz suspension returns to its original condition after the deflecting couple is removed.

(3) The torsion rod (being very small) is totally eliminated.

(4) Lamp and scale arrangement increases the accuracy of measuring deflection.

(5) The force of attraction between large and distant small spheres is negligible as they are suspended at different heights.

QUESTIONS AND PROBLEMS

5.1. (a) Define gravitational field, gravitational intensity and gravitational potential of a body.

(b) Prove that the intensity of the gravitational field is given by

$$I = -\frac{dV}{dx} = -\text{grad } V$$

where V is the gravitational potential.

5.2. What do you understand by escape and orbital velocities? Prove that the velocity of escape of a body from the earth's surface is $\sqrt{2}$ times its orbital velocity near the surface of the earth.

5.3. Derive the expressions for the gravitational potential and intensity due to a thin spherical shell at a point (a) outside the shell, (b) inside the shell.

5.4. Find the expressions for the gravitational force and potential at a point inside and outside a uniform solid sphere.

5.5. State and prove Kepler's laws of planetary motion.

5.6. Derive Newton's law of gravitation from Kepler's laws.

5.7. Define constant of gravitation. Describe Boys' method for its determination.

5.8. Describe, in brief, Cavendish method for the determination of G. What are its demerits? How are these demerits overcome in Boys' method?

5.9. Calculate the mass and mean density of the earth from the following data:

$$G = 6.7 \times 10^{-8} \text{ C.G.S. units}$$

$$g = 980 \text{ cm/sec}^2$$

Mean radius of the earth, $R = 6.4 \times 10^8$ cm

$(59.88 \times 10^{26} \text{ gm}, 5.464 \text{ gm/cc})$

5.10. Assuming the radius of the earth to be 6.4×10^8 cm and acceleration due to gravity to be 980 cm/sec², find the escape velocity of a particle from the earth's surface.

(11.2 km/sec)

5.11. Calculate the velocity for an earth satellite to orbit at a height of 200 miles above sea-level. Assume the radius of the earth to be equal to 4000 miles.

(25374 ft/sec)

5.12. The radius of the earth is 6.4×10^8 cm, its mean density 5.5 gm/c.c. and the gravitational constant 6.7×10^{-8} C.G.S. units. Calculate the earth's surface potential.

$(- 6.339 \times 10^{11}$ ergs/gm)

5.13. Assuming the earth to be a solid sphere, calculate the amount of work required to send a body of mass m from earth's surface to a height (i) $R/2$, (ii) 10 R, and (iii) 1000 R, where R is the radius of the earth. Express the result in terms of m, R and g where g is the acceleration due to gravity on earth's surface.

$$\left(\frac{1}{3} Rmg; \left(\frac{10}{11}\right) Rmg; \left(\frac{1000}{1001}\right) Rmg\right)$$

6. Simple Harmonic Oscillator

6.1. INTRODUCTION

A motion which repeats itself after a regular interval of time is called periodic motion. Periodic motion in which a particle performs to and fro motion about a fixed point is known as oscillatory motion. An oscillatory motion, in which the force is directly proportional to the displacement of the particle and is directed towards the fixed point, is called simple harmonic motion. Mathematically,

$$F \propto - x$$

or
$$F = - kx \qquad (6.1)$$

where F is the force, x is the displacement of the particle from the fixed point and k is a constant of proportionality called the force constant. The negative sign indicates that the force is opposed to the displacement.

A system performing simple harmonic-motion is called a simple harmonic oscillator. Fig. 6.1 shows a simple harmonic oscillator in the form of a body of mass m attached to an ideal spring of force constant k and free to move over a frictionless horizontal surface. Fig. 6.1b shows a simple harmonic oscillator in the equilibrium position. When the oscillator is displaced towards the right of this position (Fig. 6.1a), a restoring force ($F = - kx$) due to the spring acts towards left. The restoring force acts towards right if the oscillator is displaced towards the left of the equilibrium position. The system executes simple harmonic motion.

Applying Newton's second law of motion, the restoring force F acting on a simple harmonic oscillator is given by

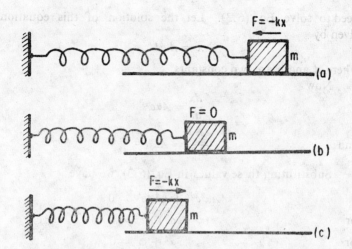

Fig. 6.1. A simple harmonic oscillator. The force exerted by the spring is shown in each case. The body slides on a frictionless table.

$$F = m \frac{d^2x}{dt^2} \qquad (6.2)$$

where m and d^2x/dt^2 denote the mass and the acceleration of the oscillator respectively.

Using Eq. (6.2) in Eq. (6.1), we write

$$m \frac{d^2x}{dt^2} = - kx$$

or

$$\frac{d^2x}{dt^2} + \frac{k}{m} x = 0$$

or

$$\frac{d^2x}{dt^2} + \omega^2 x = 0 \qquad (6.3)$$

where $\omega \left(= \sqrt{\frac{k}{m}} \right)$ is a constant to be defined later. Eq. (6.3) is

called the differential equation of a simple harmonic oscillator.

6.2. DISPLACEMENT, VELOCITY AND TIME PERIOD OF A SIMPLE HARMONIC OSCILLATOR

In order to find the displacement of a harmonic oscillator we need to solve Eq. (6.3). Let the solution of this equation be given by

$$x = Ae^{\alpha t}$$

where A and α are two constants.

Now

$$\frac{dx}{dt} = A\alpha e^{\alpha t}$$

and

$$\frac{d^2 x}{dt^2} = A\alpha^2 e^{\alpha t}$$

Substituting these values in Eq. (6.3), we have

$$A\alpha^2 e^{\alpha t} + \omega^2 Ae^{\alpha t} = 0$$

or

$$(\alpha^2 + \omega^2)\, Ae^{\alpha t} = 0$$

or

$$\alpha^2 + \omega^2 = 0$$

Hence

$$\alpha = \pm j\omega$$

Here $j\, (=\sqrt{-1})$ is the imaginary number. Therefore the solution of Eq. (6.3) is written as

$$x = A_1 e^{j\omega t} + A_2 e^{-j\omega t} \qquad (6.4)$$

where A_1 and A_2 are constants and can be determined by suitable initial conditions.

Eq. (6.4) can be further written as

$$x = A_1 (\cos \omega t + j \sin \omega t) + A_2 (\cos \omega t - j \sin \omega t)$$

or

$$x = (A_1 + A_2) \cos \omega t + j (A_1 - A_2) \sin \omega t$$

or

$$x = a \sin \phi \cos \omega t + a \cos \phi \sin \omega t$$

or

$$x = a \sin (\omega t + \phi) \qquad (6.5)$$

Here we have defined

$$A_1 + A_2 = a \sin \phi$$

and

$$j (A_1 - A_2) = a \cos \phi$$

where ϕ is a constant known as the initial phase or phase constant. The term $(\omega t + \phi)$ is called the phase of the simple harmonic oscillator.

Eq. (6.5) is the required expression for the displacement of a harmonic oscillator at any time t.

The displacement is maximum when $\sin (\omega t + \phi) = \pm 1$; that is,

$$x = \pm a$$

where a is called the amplitude of the simple harmonic oscillator.

Let

$$\varphi = \frac{\pi}{2} + \delta$$

then

$$x = a \cos (\omega t + \delta) \qquad (6.6)$$

Thus the displacement of a simple harmonic oscillator may be represented by either a sine [Eq. (6.5)] or a cosine function [Eq. (6.6)]. As they are harmonic functions, the system is called a simple harmonic oscillator. Note that the phase constants in Eqs. (6.5) and (6.6) are not the same.

The velocity of a simple harmonic oscillator is obtained by differentiating Eq. (6.5) with respect to time; that is,

$$V = \frac{dx}{dt} = a\omega \cos (\omega t + \varphi)$$

or

$$V = \omega \sqrt{a^2 - x^2} \qquad (6.7)$$

The time period of a simple harmonic oscillator is the time after which it repeats its motion. If t in Eq. (6.5) is increased by $2\pi/\omega$, we have

$$x = a \sin [\omega(t + 2\pi/\omega) + \varphi]$$

or

$$x = a \sin [2\pi + (\omega t + \varphi)]$$

or

$$x = a \sin (\omega t + \varphi)$$

This means that the motion of the harmonic oscillator repeats itself after $2\pi/\omega$ seconds. Hence the time period of the oscillator is expressed as

$$T = \frac{2\pi}{\omega} = 2\pi \sqrt{\frac{m}{k}} \qquad (6.8)$$

Frequency of a simple harmonic oscillator is the reciprocal of its time period and is given by

$$n = \frac{1}{T} = \frac{\omega}{2\pi} = \frac{1}{2\pi} \sqrt{\frac{k}{m}} \qquad (6.9)$$

Here $\omega \left(= \frac{2\pi}{T} = 2\pi n \right)$ is known as the angular frequency of the simple harmonic oscillator

6.3.   ENERGY OF A SIMPLE HARMONIC OSCILLATOR

The total energy of a simple harmonic oscillator is the sum of its kinetic and potential energies.   The kinetic energy is given by

$$K = \tfrac{1}{2}mV^2 \qquad (6.10)$$

Using Eq. (6.7) in Eq. (6.10), we write

$$K = \tfrac{1}{2}m\omega^2\,(a^2 - x^2) \qquad (6.11)$$

Potential energy of the oscillator is the work done in displacing it from $x = 0$ to $x = x$; that is,

$$U = \int^{x} - F dx = \int_{0}^{x} kx\,dx$$

or $U = \tfrac{1}{2}kx^2 \qquad (6.12)$

Adding Eqs. (6.11) and (6.12), the total energy E of the simple harmonic oscillator is expressed as

$$E = K + U$$

or $E = \tfrac{1}{2}m\omega^2\,(a^2 - x^2) + \tfrac{1}{2}kx^2$

or $E = \tfrac{1}{2}ka^2 \qquad (6.13)$

Eq. (6.13) shows that the total energy of a simple harmonic oscillator is constant and is proportional to the square of its amplitude. This is true only when no damping forces are acting on the system.

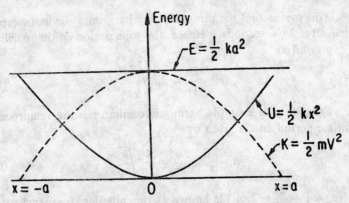

Fig. 6.2.   Variations of kinetic, potential and total energy of a simple harmonic oscillator with displacement.

At $x = 0$, the potential energy is zero $(U = 0)$ and the kinetic energy is maximum $(K = \frac{1}{2}m\omega^2 a^2)$. At $x = a$, the potential energy of the oscillator is maximum $(U = \frac{1}{2}ka^2)$ and the kinetic energy is minimum $(K = 0)$. Thus we see that the kinetic energy is maximum when potential energy is minimum and *vice versa*. But the total energy of the simple harmonic oscillator remains constant. The variations of potential energy, kinetic energy and the total energy of a simple harmonic oscillator as a function of displacement from the mean position are shown in Fig. 6.2.

6.4. EXAMPLES OF HARMONIC OSCILLATOR

1. *The Simple Pendulum*

A simple pendulum is an idealised body of point mass suspended by a flexible, weightless and inextensible string. In practice it is not possible to achieve the above mentioned conditions but a heavy metallic bob attached to a fine cotton thread can serve the purpose. When a simple pendulum is displaced from its mean position (equilibrium position), it vibrates in a vertical plane under the influence of gravity. The motion is periodic and oscillatory. For small angular displacements the motion is simple harmonic and the pendulum behaves like a simple harmonic oscillator.

Fig. 6.3 shows a simple pendulum of mass m and length l making an angle θ with the vertical. The forces acting on the system are the gravitational force mg and the tension in the string T. The force mg is resolved into two components, $mg \cos \theta$ along the radial direction and $mg \sin \theta$ along the tangent. The radial component balances the tension T which provides the necessary centripetal force to keep the bob moving along a circular arc. The tangential component $mg \sin \theta$ provides the restoring force given by

$$F = -mg \sin \theta$$

or $\qquad\qquad F = -mg\,\theta \qquad\qquad$ (θ being very small $\sin \theta \approx \theta$)

If x is the displacement of the bob along the arc, then

$$x = l\theta$$

Hence $\qquad\qquad F = -mg\dfrac{x}{l}$

or $\qquad\qquad F = -kx$

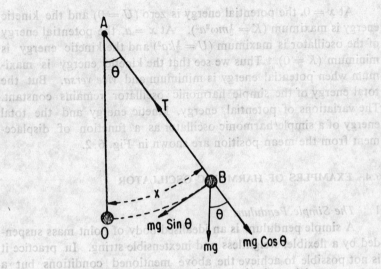

Fig. 6.3. Various forces acting on a simple pendulum.

where $k \, (= mg/l)$ is the force constant.

Therefore, the motion is simple harmonic and its time period is given by Eq. (6.8); that is,

$$T = 2\pi \sqrt{\frac{m}{k}} = 2\pi \sqrt{\frac{m}{mg/l}}$$

or
$$T = 2\pi \sqrt{\frac{l}{g}} \qquad\qquad (6.14)$$

Eq. (6.14) may be used to find the value of g, the acceleration due to gravity. Also the pendulum is useful as a time keeper as its time period is independent of its amplitude.

2. *The Torsional Pendulum*

A torsional pendulum consists of a disc suspended by a wire attached to the centre of mass of the disc. The other end of the wire is firmly fixed to a rigid support as shown in Fig. 6.4. When the disc is given a twist and released, it performs angular oscillatory motion in a horizontal plane about the equilibrium position. This is due to the restoring torque exerted by the wire on the disc tending it to return to its original condition. For small twists, the restoring torque is found to be proportional to the angular displacement and is written as

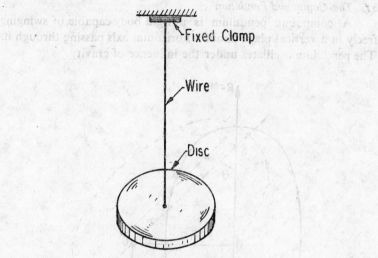

Fig. 6.4.   Torsional pendulum.

$$\tau = - C\theta \tag{6.15}$$

where C is the restoring torque per unit deflection and is known as torsional constant. The negative sign indicates that the torque is directed opposite to the angular displacement. If I is the moment of inertia of the oscillating system about its axis of rotation, then

$$\tau = I\frac{d^2\theta}{dt^2} \tag{6.16}$$

where $d^2\theta/dt^2$ is the angular acceleration. Using Eqs. (6.15) and (6.16) we get

$$I\frac{d^2\theta}{dt^2} = - C\theta$$

or

$$\frac{d^2\theta}{dt^2} + \frac{C}{I}\theta = 0 \tag{6.17}$$

Eq. (6.17) is the differential equation of an angular simple harmonic motion whose time period is given by

$$T = 2\pi\sqrt{\frac{I}{C}} \tag{6.18}$$

The principle of a torsional pendulum is employed in many instruments like galvanometers, balance wheels of watches, Cavendish balance, etc.

3. *The Compound Pendulum*

A compound pendulum is a rigid body capable of swinging freely in a vertical plane about a horizontal axis passing through it. The pendulum oscillates under the influence of gravity.

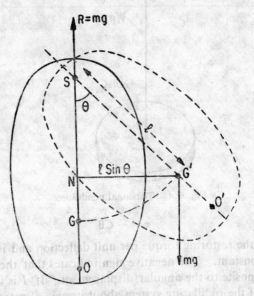

Fig. 6.5. A compound pendulum. *S* and *O* are the centres of suspension and oscillation respectively. *G* is the centre of gravity of the pendulum. Dotted section shows the displaced position. *G'* is the displaced position of centre of gravity.

The vertical section of a compound pendulum is shown in Fig. 6.5. *S* is the point of suspension and *G* is the position of centre of gravity of the pendulum in its equilibrium position. The distance between *S* and *G* is *l*. When the pendulum is displaced through an angle θ, a restoring couple is set up due to the weight (mg) of the pendulum and the reaction at the support. The moment of this couple τ (torque) is given by

$$\tau = - mgl \sin \theta \tag{6.19}$$

For small amplitudes, θ is small and $\sin \theta \approx \theta$. Thus Eq. (6.19) is rewritten as

$$\tau = - mgl\theta \tag{6.20}$$

The negative sign indicates that the torque is directed opposite to the angular displacement. Applying Newton's second law for rotatory motion, the torque τ is expressed as

$$\tau = I\frac{d^2\theta}{dt^2} \tag{6.21}$$

where I is the moment of inertia of the pendulum about a horizontal axis through S and $d^2\theta/dt^2$ is the angular acceleration.

From Eqs. (6.20) and (6.21), we have

$$I\frac{d^2\theta}{dt^2} = -mgl\theta$$

or

$$\frac{d^2\theta}{dt^2} + \frac{mgl}{I}\theta = 0 \tag{6.22}$$

Eq. (6.22) represents the differential equation of an angular simple harmonic motion whose time period is written as

$$T = 2\pi\sqrt{\frac{I}{mgl}} \tag{6.23}$$

If I_G is the moment of inertia of the pendulum about a parallel axis through G, then the theorem of parallel axes gives

$$I = I_G + ml^2$$

But

$$I_G = mK^2$$

where K is the radius of gyration of the pendulum about a horizontal axis through G.

Hence

$$I = mK^2 + ml^2$$

or

$$I = m(K^2 + l^2) \tag{6.24}$$

Substituting Eq. (6.24) in Eq. (6.23), we have

$$T = 2\pi\sqrt{\frac{K^2 + l^2}{lg}}$$

or

$$T = 2\pi\sqrt{\frac{K^2/l + l}{g}}$$

or

$$T = 2\pi\sqrt{\frac{L}{g}} \tag{6.25}$$

where $L\left(=\dfrac{K^2}{l} + l\right)$ is known as the length of an equivalent simple pendulum. A point O at a distance L from S on the line SG produced is defined as the centre of oscillation. The point of intersection of the horizontal axis through S and a vertical plane passing through G is called the point of suspension.

It can be easily shown that the centres of suspension and oscillation are interchangeable.

Fig. 6.6 shows the positions of centre of suspension S and the centre of oscillation O. The time period of the pendulum about S is given by

$$T = 2\pi \sqrt{\frac{K^2/l + l}{g}} \qquad (6.26)$$

If the pendulum is inverted and is allowed to oscillate about the centre of oscillation O, its time period T' is written as

$$T' = 2\pi \sqrt{\frac{\dfrac{K^2}{K^2/l} + \dfrac{K^2}{l}}{g}}$$

or

$$T' = 2\pi \sqrt{\frac{K^2/l + l}{g}} \qquad (6.27)$$

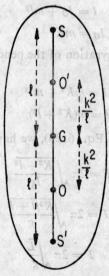

Fig. 6.6.   Centres of oscillation and suspension of a compound pendulum.

From Eqs. (6.26) and (6.27), we have

$$T = T'$$

Thus the time periods of the pendulum about the centre of suspension and the centre of oscillation are the same; that is, the centres of suspension and oscillation are interchangeable. In fact there are four points (Fig. 6.6) collinear with the centre of gravity G about which the time period of the pendulum is the same. The principle that the centres of suspension and oscillation are interchangeable is employed to determine the value of acceleration due to gravity by a compound pendulum. For this the compound pendulum may be used in two ways—as a bar pendulum and as a Kater's pendulum. We locate two points in the pendulum at unequal distances on either side of its centre of gravity and collinear with it such that the time periods about them are equal. One of these points will be the centre of suspension and the other will be the corresponding point of oscillation. The distance between them gives the length of an equivalent simple pendulum. The value of g can be determined by using Eq. (6.25).

Let us study the variation of time period of a compound pendulum with the distance between its centre of gravity and the point of suspension.

From Eq. (6.26), we have

$$T^2 = \frac{4\pi^2}{g} \left(\frac{K^2}{l} + l \right) \tag{6.28}$$

Differentiating Eq. (6.28) with respect to l, we get

$$2T \frac{dT}{dl} = \frac{4\pi^2}{g} \left(\frac{-K^2}{l^2} + 1 \right)$$

or

$$\frac{dT}{dl} = \frac{\pi \left(\dfrac{-K^2}{l^2} + 1 \right)}{\sqrt{g \left(\dfrac{K^2}{l} + l \right)}} \tag{6.29}$$

Eq. (6.29) gives the variation of T with l. For minimum and maximum values of the time period $dT/dl = 0$; that is,

$$-\frac{K^2}{l^2} + 1 = 0$$

or

$$l = \pm K \tag{6.30}$$

The value $l = -K$ is not permissible as in that case the time period becomes imaginary. Also, the time period of the compound pendulum is maximum when $l = 0$ ($T \to \infty$ as $l \to 0$, Eq. (6.26)). Thus the time period of the compound pendulum will be minimum when $l = K$; that is, when the distance between the centre of suspension and the centre of gravity is equal to the radius of gyration of the pendulum about a parallel axis through the centre of gravity. Thus

$$T_{\min} = 2\pi \sqrt{\frac{2K}{g}}$$

4. *Kater's Pendulum*

A Kater's pendulum consists of a metal bar having two adjustable knife edges A and B on either side of its centre of gravity (Fig. 6.7). W_1, W_2 and w are three movable weights. W_1 and W_2 are identical in shape but W_1 is made of wood while W_2 is a

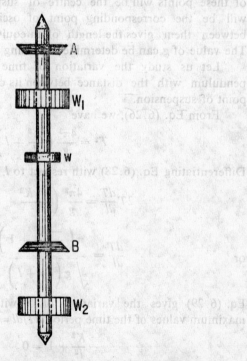

Fig. 6.7.   A Kater's pendulum. A and B are knife edges. W_1 is wooden weight. W_2 is a heavy metallic weight. w is a small metallic weight.

metallic weight. w is a small metallic weight used for finer adjustment.

The pendulum can be suspended on either of the two knife edges A and B and its time period determined. By suitably adjusting W_1, W_2 and w (particulary W_2) the time periods about A and B are made exactly equal. Then one of the knife edges will be the centre of suspension and the other the corresponding centre of oscillation. The distance between A and B will give the length of equivalent simple pendulum. The value of g can be determined by using the formula

$$T = 2\pi \sqrt{\frac{L}{g}}$$

Usually it is very difficult to make the time periods equal about A and B. Bessel showed that it was not necessary to make the time periods equal, and that g could be calculated even when we made the time periods nearly equal about A and B.

Let T_1 and T_2 be the time periods of the pendulum about A and B respectively. T_1 is nearly equal to T_2. Using Eq. (6.26) we have

$$T_1 = 2\pi \sqrt{\frac{K^2 + l_1^2}{l_1 g}} \tag{6.31}$$

and

$$T_2 = 2\pi \sqrt{\frac{K^2 + l_2^2}{l_2 g}} \tag{6.32}$$

where l_1 and l_2, respectively, are the distances of knife edges A and B from the centre of gravity of the pendulum.

Squaring Eq. (6.31), we get

$$T_1^2 = 4\pi^2 \frac{(K^2 + l_1^2)}{l_1 g}$$

or

$$l_1 T_1^2 = \frac{4\pi^2}{g} (K^2 + l_1^2) \tag{6.33}$$

Similarly, from Eq. (6.32), we have

$$l_2 T_2^2 = \frac{4\pi^2}{g} (K^2 + l_2^2) \tag{6.34}$$

Subtracting Eq. (6.34) from Eq. (6.33), we write

$$l_1 T_1^2 - l_2 T_2^2 = \frac{4\pi^2}{g} (l_1^2 - l_2^2)$$

or
$$\frac{4\pi^2}{g} = \frac{l_1 T_1^2 - l_2 T_2^2}{l_1^2 - l_2^2} \qquad (6.35)$$

Using partial fractions, Eq. (6.35) can be written as

$$\frac{4\pi^2}{g} = \frac{l_1 T_1^2 - l_2 T_2^2}{(l_1 + l_2)(l_1 - l_2)} = \frac{A}{l_1 + l_2} + \frac{B}{l_1 - l_2} \qquad (6.36)$$

or
$$\frac{l_1 T_1^2 - l_2 T_2^2}{(l_1 + l_2)(l_1 - l_2)} = \frac{A(l_1 - l_2) + B(l_1 + l_2)}{(l_1 + l_2)(l_1 - l_2)}$$

or
$$\frac{l_1 T_1^2 - l_2 T_2^2}{(l_1 + l_2)(l_1 - l_2)} = \frac{l_1(A + B) - l_2(A - B)}{(l_1 + l_2)(l_1 - l_2)}$$

Comparing the coefficients of l_1 and l_2 on both sides, we get

$$A + B = T_1^2 \text{ and } A - B = T_2^2$$

Therefore
$$A = \frac{T_1^2 + T_2^2}{2}$$

and
$$B = \frac{T_1^2 - T_2^2}{2} \qquad (6.37)$$

Substituting the values of A and B from Eq. (6.37) in Eq. (6.36), we have

$$\frac{4\pi^2}{g} = \frac{T_1^2 + T_2^2}{2(l_1 + l_2)} + \frac{T_1^2 - T_2^2}{2(l_1 - l_2)} \qquad (6.38)$$

The distances l_1 and l_2 can be determined by finding the position of centre of gravity of the pendulum. Knowing T_1 and T_2, the value of g can be calculated from Eq. (6.38). As there is an element of uncertainty in the determination of the centre of gravity, we make T_1 to be nearly equal to T_2. In that case the second term in Eq. (6.38) becomes negligible (l_1 is different from l_2). The distance $l_1 + l_2$ is the distance between the knife edges and can be measured accurately. Therefore the value of g can be determined by using the formula

$$\frac{4\pi^2}{g} = \frac{T_1^2 + T_2^2}{2(l_1 + l_2)} \qquad (6.39)$$

For an accurate determination of g by a Kater's pendulum, following corrections are applied:

(a) *Finite Amplitude.* The expression of the time period of a Kater's pendulum is true only for vanishingly small angular amplitudes. It is then only that we can take $\sin \theta \approx \theta$ and the motion of the pendulum is simple harmonic. But in actual practice this is not true and the amplitude has a finite magnitude as a result of which the observed time period is greater than the ideal time period for an infinitely small amplitude. Therefore a correction for this has to be applied to get an accurate value of g.

If during an experiment, the amplitude falls from α_1 to α_2 where each is small, it can be proved that the corrected time period T_C is given by

$$T_C = T\left(1 - \frac{\alpha_1 \alpha_2}{16}\right) \qquad (6.40)$$

(b) *Air Correction.* Buoyancy of air affects the restoring couple as it exerts an upward thrust on the pendulum. This error is quite complex and can be minimised either by performing the experiment in an evacuated chamber or by higher mathematical calculations. As the pendulum vibrates it drags some air along with it resulting in the increase in its effective moment of inertia. According to Bessel this effect is eliminated if the pendulum is made symmetrical about its geometric centre.

(c) *Air Damping.* In the theory of Kater's pendulum the effect of air damping has not been taken into account. The air, due to its viscosity, opposes the motion of the pendulum and so the observed time period is greater than its ideal value. This can be eliminated by taking into account the damping effect in the differential equation of the motion of the pendulum.

(d) *Curvature of the Knife-edges.* When the knife-edges are not sharp, the effective length of the pendulum will change causing a change in the time period. The effect of the knife-edge curvature may be avoided by having plane bearings on the pendulum and a fixed knife-edge on the support. The knife-edge is ground to a fairly sharp-edge and the plane bearings are made accurately flat and are always replaced in the same position on the knife-edge.

(e) *Yielding of Support.* If the support is not rigid, it will yield resulting in a change of the time period of the pendulum. A non-rigid support is forced to oscillate coperiodically with the pendulum. This motion may be resolved into vertical and horizontal components. Of these, the latter has much greater effect on the period and may become an extremely disturbing factor. It is thus necessary to arrange that the support is fixed rigidly, particularly in a lateral direction, and to ensure that no cumulative resonance effect is permitted.

(*f*) *Temperature.* The change of temperature will cause a change in the length of the pendulum resulting in a variation of the time period.

(*g*) *Rotation of the Earth.* As the earth is rotating about its polar axis, the real value of *g* is obtained by combining its observed value with the centripetal acceleration at the place.

5. *Mass on a Weightless Spring*

    Consider the motion of a weightless spring *S* one end of which is fixed to a rigid support *R* and the other end carries a mass *m* (Fig. 6.8). When *m* is pulled a little and released, the system

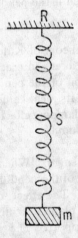

Fig. 6.8. A mass *m* suspended by a weightless spring *S* fixed to a rigid support *R*.

performs simple harmonic motion and the restoring force *F* is given by

$$F = -kx$$

where *k* is spring constant and *x* is the displacement of *m* from its mean position (equilibrium position). According to Newton's second law of motion $F = m \dfrac{d^2x}{dt^2}$, where $\dfrac{d^2x}{dt^2}$ is the acceleration of mass m. Hence the equation of motion of the mass m is given by

$$m\frac{d^2x}{dt^2} + kx = 0$$

or

$$\frac{d^2x}{dt^2} + \frac{k}{m}\,x = 0$$

This is the equation of a simple harmonic motion whose time period T is expressed as

$$T = 2\pi\,\sqrt{\frac{m}{k}} \qquad (6.41)$$

and the displacement x is given by

$$x = a \sin(\omega t + \phi)$$

6. Oscillation of a Magnet

When a freely suspended magnet of length $2l$ and pole strength m is allowed to oscillate with a small angular amplitude θ in earth's magnetic field H, it performs simple harmonic motion. The

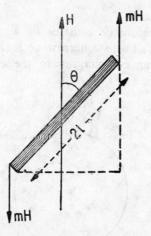

Fig. 6.9. A freely suspended magnet oscillating in a uniform field H. m is the pole strength.

restoring couple (torque τ) on the magnet due to earth's magnetic field is given by

$$\tau = -\,2ml\,H\sin\theta$$

or $$\tau = -MH\theta \qquad \text{(For small } \theta, \sin \theta \approx \theta)$$
$$(6.42)$$

where $M (= 2ml)$ is the magnetic moment of the magnet. Applying Newton's second law of rotatory motion

$$\tau = I\frac{d^2\theta}{dt^2} \qquad (6.43)$$

where I is the moment of inertia of the magnet about the point of suspension and $\frac{d^2\theta}{dt^2}$ is its angular acceleration.

Combining Eqs. (6.42) and (6.43), we get

$$I\frac{d^2\theta}{dt^2} + MH\theta = 0$$

or $$\frac{d^2\theta}{dt^2} + \frac{MH}{I}\theta = 0 \qquad (6.44)$$

Eq. (6.44) represents the differential equation of an angular simple harmonic motion whose time period T is given by

$$T = 2\pi \sqrt{\frac{I}{MH}} \qquad (6.45)$$

7. Helmholtz Resonator

A Helmholtz resonator consists of a round or cylindrical shaped metallic vessel with a narrow neck and a small hole at its other end which can communicate to the ear (Fig. 6.10). The

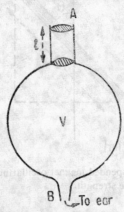

Fig. 6.10. A Helmholtz resonator. A and l are the area and length of its neck. V is its volume. B is a small hole which can communicate to the ear.

resonator can be excited through its neck, the air in which functions like a piston which compresses and rarifies the gas inside the vessel. Let A and l be the area and length of the neck of the resonator respectively. If the air within the neck moves through a small distance x, the change in the volume of the gas in the vessel is $v = xA$. If V is the volume of the resonator, E is the bulk modulus of elasticity, then the increase in pressure p in the resonator is given by

$$E = -\frac{p}{\dfrac{v}{V}}$$

or

$$p = -\frac{Ev}{V} = -\frac{ExA}{V}$$

Hence the force exerted on the air in the neck of the resonator is written as

$$F = pA = -\frac{EA^2x}{V} \qquad (6.46)$$

If ρ is the density of air, the mass of the air in the neck of the resonator is ρlA. Applying Newton's second law of motion the force acting on the air is written as

$$F = \rho lA \frac{d^2x}{dt^2} \qquad (6.47)$$

where d^2x/dt^2 is the acceleration of the air in the neck.

Combining Eqs. (6.46) and (6.47), we get

$$\rho lA \frac{d^2x}{dt^2} + \frac{ExA^2}{V} = 0$$

or

$$\frac{d^2x}{dt^2} + \frac{EA}{l\rho V}x = 0 \qquad (6.48)$$

Eq. (6.48) is the differential equation of a simple harmonic motion whose time period is given by

$$T = 2\pi \sqrt{\frac{l\rho V}{EA}} \qquad (6.49)$$

If V_a is the velocity of sound in air, then

$$V_a = \sqrt{\frac{E}{\rho}} \quad \text{(Newton's formula)}$$

Hence $$T = \frac{2\pi}{V_a} \sqrt{\frac{lV}{A}}$$ (6.50)

and the frequency n of the resonator is written as

$$n = \frac{1}{T} = \frac{V_a}{2\pi} \sqrt{\frac{A}{Vl}}$$ (6.51)

n is the natural frequency of the resonator. When the frequency of the incident sound wave is equal to the natural frequency of the resonator, it will resonate and the resonance can be detected by putting the ear at B.

QUESTIONS AND PROBLEM

6.1. Define a simple harmonic oscillator. Discuss some of its important applications.

6.2. Derive and solve the differential equation of a simple harmonic oscillator and hence find the expressions for its velocity and time period.

6.3. Define phase, amplitude and angular frequency of a simple harmonic oscillator. Prove that the mechanical energy of a simple harmonic oscillator is always constant and is proportional to the square of its amplitude.

6.4. Derive an expression for the energy of a simple harmonic oscillator. Show tha t its average kinetic energy is equal to its average potential energy.

$$\left[\text{Hint. } K_{av} = \frac{1}{T} \int_0^T K dt, \quad U_{av} = \frac{1}{T} \int_0^T U dt \right]$$

6.5. Derive expressions for the differential equations of the following harmonic oscillators: (1) Simple pendulum, (2) Compound pendulum, (3) Mass on a spring, (4) Torsional pendulum, (5) Helmholtz resonator and hence write down the formulae for their time period and frequency.

6.6. Make a comparative study of a simple and compound pendulum.

6.7. How will you determine the value of 'g' with a compound pendulum? Prove that there are four points collinear with the centre of gravity of a compound pendulum about which its time period is the same.

6.8. (a) Define centre of suspension and centre of oscillation of a compound pendulum. Show that they are interchangeable.

(b) Derive the expressions for the maximum and minimum time periods of a compound pendulum.

6.9. What is Kater's pendulum? Derive Bessel's formula for the determination of 'g'. Mention and discuss the various corrections necessary for the accurate determination of 'g' with a Kater's pendulum.

6.10. A particle executes simple harmonic motion whose time period is 8 seconds and amplitude is 4 cm. Find the velocity and acceleration and also their maximum values when the particle is 2 cm from the mean position.

$$\begin{pmatrix} 2.72 \text{ cm/sec, } 3.14 \text{ cm/sec,} \\ -1.23 \text{ cm/sec}^2, -2.46 \text{ cm/sec}^2 \end{pmatrix}$$

6.11. A particle is performing simple harmonic motion. Its velocities at distances 1 metre and 2 metres from the mean position are 3 metres/sec and 2 metres/sec respectively. Find the amplitude and the time period of the motion.

(2.53 metres, 4.86 sec)

6.12. In a simple harmonic motion, maximum velocity is 100 cm/sec and the maximum acceleration is 157 cm/sec.^{2} Calculate the time period of the motion. (4 sec)

6.13. A particle is executing simple harmonic motion in a straight line. Its velocities at distances x_1 and x_2 from the mean position are v_1 and v_2 respectively. Prove that the time period of the motion is

$$T - 2\pi \left[\frac{x_2{}^2 - x_1{}^2}{v_1{}^2 - v_2{}^2} \right]^{1/2}$$

[Hint. $v_1 = \omega\sqrt{a^2 - x_1{}^2}$, $v_2 = \omega\sqrt{a^2 - x_2{}^2}$]

6.14. Assume the earth to be homogeneous sphere of radius R and density ρ. Let a tunnel be bored through the earth through its centre. Prove that a cricket ball dropped into the tunnel will execute a simple harmonic motion whose time period is given by

$$T = 2\pi \sqrt{\frac{R}{g}}$$

where g is the acceleration due to gravity.

$$\left[\text{Hint.} \quad \begin{array}{l} \text{Value of } g \text{ at a distance } x \text{ from the centre of the} \\ \text{earth is } g' = (g/R)x \end{array}\right]$$

6.15. A disc of metal of radius R with its plane vertical can be made to swing about a horizontal axis passing through any one of a series of holes bored along a diameter. Show that the minimum time period of oscillation is given by

$$T = 2\pi \sqrt{\frac{1.414R}{g}}$$

6.16. A compound pendulum is formed by suspending a heavy ring of radius 490.5 cm from a string. What is the length of the string? When is the period minimum? Find this period. $g = 981$ cm/sec².

(490.5 cm, 6.28 sec)

6.17. A 800 gm mass extends a spring 4 cm from its unstretched position. The mass is replaced by a body of 50 gm mass. If this mass is pulled and then released, find the period of oscillation. $g = 980$ cm/sec².

(0.1 sec)

6.18. A solid sphere whose mass is 4 kg and radius is 0.05 metre is suspended by a wire. If the torque required to twist the wire is 4×10^{-3} Newton-metre/radian, find the period of oscillation.

(6.28 sec)

6.19. A vessel has a neck of 2 cm diameter and 2.5 cm length. If the capacity of the vesssl is 2 litres, calculate the wave length to which the vessel will resonate.

(250.7 cm)

7. Elasticity and Viscosity

7.1. ELASTICITY

Introduction

A body in which there is no relative displacement of its particles when an external force is applied to it, is called a rigid body. Such a body always maintains its shape and there is no change in its length, volume or shape under the action of external or deforming forces. In actual practice there is no body which is perfectly rigid, and there is always a change in its length, volume or shape due to the deforming forces. The body may or may not regain its original condition after the deforming forces are removed. A body which regains its original condition completely after the removal of external forces, is called perfecty elastic. On the other hand, if it remains deformed, the body is said to be perfectly plastic. In fact there is no body which is either perfectly elastic or perfectly plastic. All bodies lie between these two limits. Nearest approach to a perfectly elastic body is a quartz fibre while putty may be regarded as a nearly perfectly plastic body. The subject which deals with the behaviour of bodies in relation to the above property is called elasticity.

The property by virtue of which a body regains its original condi tion after the deforming forces are removed is called elasticity.

7.2. STRESS, STRAIN AND HOOKE'S LAW

Stress. According to Newton's third law of motion, to every action there is an equal and opposite reaction. When an external force is applied to a body there develops an equal and opposite reaction which tries to restore the original state of the body. This restoring force set up inside the body per unit area is called stress. Within the elastic limit stress is measured by the applied force per unit area of the body. Mathematically,

$$\text{Stress} = \frac{F}{A} \qquad (7.1)$$

where F is the force acting on the area A.

Depending on the type of force applied to a body, stress is divided into three types. If the force is normal and longitudinal, the stress is called longitudinal stress. A normal force applied throughout the volume of a body gives rise to a volume stress. If the force applied is tangential, the stress is known as tangential stress. In C.G.S. and M.K.S. systems, the units of stress are dynes/cm² and Newtons/m² respectively.

Strain. A body is said to be under strain when the external forces produce a change in its length, volume or shape.

Strain is of three types depending on the type of stress acting on the body. For a longitudinal stress, the strain produced is known as longitudinal strain. It is measured by the change in length per unit original length of a body. A volume stress causes a volume strain which is measured by the change in volume per unit original volume of a body. When the stress is a tangential stress, the strain produced is called a tangential strain or shearing strain. In this case there is a change in the shape of the body without change in volume. A tangential strain is measured by the angle through which a line originally perpendicular to the fixed face of a body is turned. Strain has no dimensions as it is the ratio of two similar quantities.

Hooke's Law. Hooke's law was established by Robert Hooke in 1678. It is a relation between stress and strain. *Hooke's law states that, within the elastic limit, the stress varies directly as the strain.* Mathematically,

$$\text{Stress} \propto \text{Strain}$$

or
$$E = \frac{\text{Stress}}{\text{Strain}} \qquad (7.2)$$

where E is a constant called the modulus of elasticity. Its value depends on the nature and character of the body and is independent of the magnitude of the stress or the strain. The units of the modulus of elasticity are the same as those of the stress.

7.3. THE ELASTIC CONSTANTS

In the previous section we have seen that there are three types of stresses and strains. Corresponding to these three types, we have three modulii of elasticity which have been discussed below.

1. *Young's Modulus*

Within the elastic limit, the ratio of the longitudinal stress to the corresponding longitudinal strain is known as Young's modulus. It is denoted by the symbol Y. Mathematically,

$$Y = \frac{\text{longitudinal stress}}{\text{longitudinal strain}}$$

or

$$Y = \frac{F/A}{l/L} = \frac{FL}{Al} \qquad (7.3)$$

where F, A, l and L are, respectively, the force applied, the area, the change in length and the original length of the body.

Figure 7.1 shows a curve between the load applied to a body and the corresponding extension produced in it. Between O and A

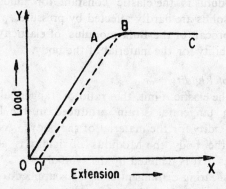

Fig. 7.1. Graph between load and extension. OA is the elastic region. A is the elastic limit. B is yielding point. BC is breaking region.

stress is proportional to strain and Hooke's law is obeyed. This is called the elastic region. Point A is called the elastic limit. The region AB is partly elastic and partly plastic. If the load is removed at this stage there is a permanent extension OO' in the body. At point B the body is permanently deformed and this point is known as the yielding point. Beyond B the deformation is rapid and finally the body breaks. The stress at this stage is known as breaking stress.

2. Bulk Modulus

Within the elastic limit, the ratio of the volume stress to the corresponding volume strain produced in the body is called Bulk modulus of elasticity. It is denoted by the symbol K. Mathematically it is given by

$$K = \frac{\text{Volume stress}}{\text{Volume strain}}$$

or

$$K = -\frac{F/A}{\dfrac{v}{V}}$$

or

$$K = -\frac{pV}{v} \qquad\qquad (7.4)$$

where $p(= F/A)$ is the pressure applied and v and V are, respectively, the change in volume and the original volume of the body. The negative sign indicates that the volume decreases with the increase in pressure and vice versa.

Bulk modulus is the elastic constant for fluids (liquids and gases) as the solids are hardly affected by pressure.

The reciprocal of the Bulk modulus of elasticity is known as the compressibility for the material of the body.

3. Modulus of Rigidity

Within the elastic limit, the ratio of tangential stress to the corresponding tangential strain produced in the body is called Modulus of rigidity for the material of the body. As it determines the shape of the body, the Modulus of rigidity is also known as Shear modulus. It is denoted by the symbol 'η'.

When a shearing or tangential stress is applied to a body, there is a change in its shape without change in its volume. Consider a cube $ABCDHEFG$ of side L. The lower edge CD of the cube is fixed and a tangential force F is applied to the upper face in the direction as shown in Fig. 7.2. As a result a force of reaction, equal in magnitude but opposite in direction to F, will act at the lower face of the cube. These two equal and opposite forces will constitute a couple which will displace the upper face through an angle θ ($EABF$ to $E'A'B'F'$). The angle θ is called the angle of shear or the shearing strain. For small value of the angle of shear

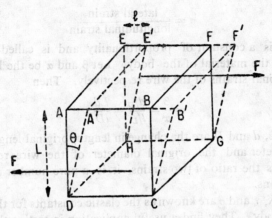

Fig. 7.2. A shearing force F applied to a cube $ABCDHEFG$. Face $CDHG$ is fixed. Cube is sheared to $A'B'CDHE'F'G$ through an angle θ

$$\theta = \tan\theta = \frac{AA'}{AD} = \frac{l}{L} \qquad (7.5)$$

Hence the Modulus of rigidity is given by

$$\eta = \frac{\text{Shearing stress}}{\text{Shearing strain}}$$

or $\qquad\qquad \eta = \frac{F/A}{\theta} = \frac{T}{\theta} \qquad\qquad (7.6)$

where $T(= F/A)$ is the shearing stress. A is the area of the cube. Units of η are the same as those of stress.

4. *Poisson's Ratio.* When a longitudinal force is applied to a wire there is an increase in its length followed by a contraction in its diameter. The increase in length and the contraction are at right angles to each other. This fact is quite general and applies also to bodies other than the wire. The ratio of the change in lateral dimension of a body to its original lateral dimension perpendicular to the external force is called lateral strain.

Within the elastic limit, the lateral strain varies directly as the longitudinal strain.

Mathematically,

lateral strain $\propto$ longitudinal strain

or $$\sigma = \frac{\text{lateral strain}}{\text{longitudinal strain}}$$

where σ is a constant of proportionality and is called Poisson's ratio for the material of the body. Let β and α be the lateral and longitudinal strains of the wire respectively. Then

$$\sigma = \frac{\beta}{\alpha} = \frac{d/D}{l/L} = \frac{Ld}{lD}. \qquad (7.7)$$

Here l, L, d and D are the change in length, original length, change in diameter and the original diameter of the wire respectively. Since σ is the ratio of two strains, it is a pure number and has no dimensions.

Y, K, η and σ are known as the elastic constants for the material of the body. They find a useful application in testing the strength of materials. In the forthcoming section we establish the relations between them.

7.4. RELATIONS BETWEEN ELASTIC CONSTANTS

Before deriving the various relations between elastic constants, we prove that a shearing stress is equivalent to an equal extension stress and an equal compression stress at right angles to each other.

Figure 7.3 shows the front view $ABCD$ of a cube of a side L. The lower face CD is fixed and a tangential force F is applied on the face AB. The force on AB will give rise to an equal and opposite force of reaction on the fixed face CD. The two forces will constitute a couple which will tend to rotate the cube in the clockwise direction. Since the cube does not rotate (CD being fixed) there must be an equal and opposite couple (anticlockwise) formed by the forces acting along CB and AD caused by the fixed face CD.

Hence a tangential force acting on the face AB gives rise to equal tangential forces along all other faces in the directions indicated in Fig. 7.3. Resultants of the forces at A and C are acting inward and their magnitudes are equal ($F\sqrt{2}$). Resultants of the forces at B and D are acting outward and their magnitudes are also equal to $F\sqrt{2}$ each. Thus outward pulls at B and D will result in the extension of BD while the inward pulls at A and C will cause a contraction in AC.

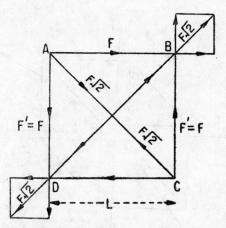

Fig. 7.3. *ABCD* is the front view of a cube. *CD* is fixed. A tangential force *F* gives rise to forces of extension and compression.

If a plane, perpendicular to the plane of paper, is drawn through *AC*, then its area is given by

$$L \times AC = L \times L\sqrt{2} = L^2\sqrt{2}$$

Hence the compression stress on $AC = \dfrac{F\sqrt{2}}{L^2\sqrt{2}} = \dfrac{F}{L^2}$

Similarly we can prove that the extension stress on

$$BD = \frac{F\sqrt{2}}{L^2\sqrt{2}} = \frac{F}{L^2}$$

Also, the shearing stress on face $AB = F/L^2$.

Therefore a shearing stress on *AB* is equal to an equal extension stress (on *BD*) and equal compression stress (on *AC*) at right angles to each other.

1. *Relation between Y, K and σ*

Consider a cube *ABCDEFGH* of side *L*. Let a normal uniform volume stress *T* act on each face in the outward direction as shown in Fig. 7.4. If α and β respectively are the longitudinal and lateral strains per unit stress, then final sides of the cube are given by

$$AB = L + LT\alpha - LT\beta - LT\beta$$

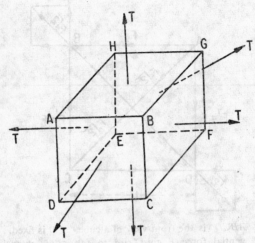

Fig. 7.4. A cube $ABCDEFGH$ of side L under the action of a normal and outward uniform volume stress T.

or $$AB = L[1 + T(\alpha - 2\beta)]$$

Here $LT\alpha$ is the increase in length of AB due to α and $2LT\beta$ is the contraction along AB due to T along BG and BC.

Similarly
$$BG = L[1 + T(\alpha - 2\beta)]$$

and $$BC = L[1 + T(\alpha - 2\beta)]$$

Hence the final volume of the cube is given by

$$AB \times BG \times BC = L^3[1 + T(\alpha - 2\beta)]^3$$

or final volume of the cube $= L^3[1 + 3T(\alpha - 2\beta)]$

(using Binomial theorem)

Now initial volume of the cube $= L^3$

Therefore the change in the volume $= 3L^3T(\alpha - 2\beta)$

Thus volume strain $= \dfrac{3L^3T(\alpha - 2\beta)}{L^3}$

or volume strain $= 3T(\alpha - 2\beta)$

Hence Bulk modulus $K = \dfrac{T}{3T(\alpha - 2\beta)} = \dfrac{1}{3(\alpha - 2\beta)}$

or $$K = \frac{1/\alpha}{3(1 - 2\beta/\alpha)} = \frac{Y}{3(1 - 2\sigma)} \qquad (7.8)$$

where $Y(= 1/\alpha)$ and $\sigma(= \beta/\alpha)$ are the Young's modulus and Poisson's ratio of the material of the cube respectively.

2.   *Relation between Y, η and σ*

$ABCD$ represents the front face of a cube of side L. CD is fixed and a tangential force F is applied to the face AB. A moves to A' and B moves to B' such that $AA' = BB' = l$. The face $ABCD$ is sheared to $A'B'CD$ through an angle θ as shown in Fig. 7.5.

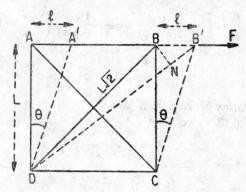

Fig. 7.5.   Front face $ABCD$ of a cube of side L. F is a shearing force on AB. θ is the angle of shear. CD is fixed.

Using Eq. (7.6), we have

$$\eta = \frac{T}{\theta}$$

where η is the modulus of rigidity, $T(= F/L^2)$ is the shearing stress and $\theta(= l/L)$ is the shearing strain. If α and β are the longitudinal and lateral strains per unit stress respectively, then the diagonal DB suffers a total extension given by

$$DB\,T\alpha + DB\,T\beta = DB\,T(\alpha + \beta)$$

or

$$NB' = DB\,T(\alpha + \beta) \qquad (7.9)$$

In Fig. 7.5, BN is perpendicular to DB'.

Now $\qquad DB' = DN + NB' = DB + NB'$

$$(DB \approx DN \text{ as } \theta \text{ is small})$$

or $\qquad DB' - DB = NB'$

This means NB' is the total extension in DB. Since θ is small $\angle AB'C$ is nearly a right angle. Hence $\angle BB'N = 45°$.

Thus

$$NB' = l \cos 45° = \frac{l}{\sqrt{2}} \qquad (7.10)$$

Using Eqs. (7.9) and (7.10), we get

$$\frac{l}{\sqrt{2}} = DBT(\alpha + \beta) = L\sqrt{2}\,T(\alpha + \beta) \quad (DB = L\sqrt{2})$$

or

$$\frac{l}{L} = 2T(\alpha + \beta)$$

or

$$\frac{T}{l/L} = \frac{T}{\theta} = \frac{1}{2(\alpha + \beta)} = \frac{1/\alpha}{2(1 + \beta/\alpha)} \left(\because \frac{l}{L} = \theta\right)$$

Hence

$$\eta = \frac{T}{\theta} = \frac{Y}{2(1 + \sigma)} \qquad (7.11)$$

3. *Relation between Y, K and η*
From Eq. (7.8) we have

$$K = \frac{Y}{3(1 - 2\sigma)}$$

or

$$(1 - 2\sigma) = \frac{Y}{3K} \qquad (7.12)$$

Also from Eq. (7.11), we get

$$2(1 + \sigma) = \frac{Y}{\eta} \qquad (7.13)$$

Adding Eqs. (7.12) and (7.13), we write

$$3 = \frac{Y}{3K} + \frac{Y}{\eta}$$

or

$$Y = \frac{9\eta K}{3K + \eta} \qquad (7.14)$$

4. *Relation between K, η and σ*
From Eqs. (7.8) and (7.11), we get

$$Y = 3K(1 - 2\sigma) = 2\eta(1 + \sigma)$$

or

$$3K - 6K\sigma = 2\eta + 2\eta\sigma$$

or

$$2\eta\sigma + 6K\sigma = 3K - 2\eta$$

or

$$\sigma(2\eta + 6K) = 3K - 2\eta$$

Hence

$$\sigma = \frac{3K - 2\eta}{2\eta + 6K} \qquad (7.15)$$

Using Eqs. (7.8) and (7.11) we can find the limiting values of σ. These equations give

$$3K(1 - 2\sigma) = 2\eta(1 + \sigma) \qquad (7.16)$$

If σ is positive, the right side of Eq. (7.16) is positive. Hence the left side should also be positive. This is possible only when

$$1 - 2\sigma > 0$$

or $\sigma < 0.5$

If σ is negative, the left side of Eq. (7.16) is positive. Hence the right side of this equation should also be positive. This happens when

$$1 + \sigma > 0$$

or $\sigma > -1.$

Hence the Poisson's ratio lies between 0.5 and -1; that is,

$$-1 < \sigma < 0.5$$

7.5. TWISTING COUPLE ON A CYLINDRICAL ROD OR WIRE

Fig. 7.6a shows a cylindrical wire of length l and radius r. The upper end of the wire is fixed and a twisting couple is applied at the lower end in a plane perpendicular to the length of the wire. Due to elasticity a restoring couple is set up in the wire. The restoring couple opposes the twisting couple and in the equilibrium position the two balance each other. Our object is to calculate this couple.

Imagine the cylindrical wire to be divided into a large number of thin coaxial cylindrical shells, and consider one such shell of radius x, thickness dx and length l. As the wire is twisted, a line AB on the shell and parallel to the axis OO' of the wire, is sheared to AB' through an angle ϕ. Fig. 7.6b shows the lower section of the wire. Now cut the cylindrical shell along AB before twisting and flatten it out. It will form a rectangle $ABCD$ whose sides are l and $2\pi x$ as indicated in Fig. 7.6c. After twisting $ABCD$ is sheared to $AB'C'D$. Hence ϕ is the angle of shear. Angle θ $(= \angle BO'B')$ is called the angle of twist at the free end of the cylindrical shell. From Fig. 7.6a, we have

$$l\phi = x\theta$$

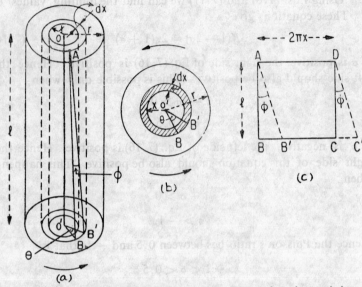

Fig. 7.6a. A cylindrical wire fixed at the upper end and a twisting
couple applied at the other end. OO' is axis of the wire.
AB is a line on one of the coaxial shells of the wire and is
parallel to OO'. AB is sheared to AB'. ϕ is the angle of
shear. θ is the angle of twist.

b.   The lower section of the wire. x is the radius of a coaxial shell
of thickness dx.

c.   Cylindrical shell cut along AB and flattened out.

or
$$\phi = \frac{x\theta}{l} \qquad (7.17)$$

If F is the shearing force acting on the base of the shell, the
Modulus of rigidity of the material of the wire is given by

$$\eta = \frac{F/\text{area of the base of the shell}}{\phi}$$

or
$$\eta = \frac{F/2\pi x dx}{\phi}$$

Hence $F = 2\pi x \eta\, \phi\, dx$ (7.18)

Substituting the value of ϕ from Eq. (7.17) in Eq. (7.18), we get

$$F = \frac{2\pi\eta\theta}{l} x^2 dx$$

Thus the twisting couple '$d\tau$' on the cylindrical shell of
radius x is written as

$$d\tau = F \cdot x = \frac{2\pi\eta\theta}{l} x^3 dx \qquad (7.19)$$

Total twisting couple acting on the wire is obtained by integrating Eq. (7.19) between the limits $x = 0$ and $x = r$; that is,

$$\tau = \int_0^r d\tau = \int_0^r \frac{2\pi\eta\theta}{l} x^3 dx$$

or

$$\tau = \frac{2\pi\eta\theta}{l} \left[\frac{x^4}{4}\right]_0^r$$

or

$$\tau = \frac{2\pi\eta r^4\theta}{4l}$$

or

$$\tau = \frac{\pi\eta r^4}{2l} \theta \qquad (7.20)$$

Therefore the twisting couple per unit twist is $\pi\eta r^4/2l$. As the twisting couple, in the equilibrium position, is equal to the restoring couple, Eq. (7.20) also represents the restoring couple. Hence the restoring couple per unit twist C is given by

$$C = \frac{\pi\eta r^4}{2l} \qquad (7.21)$$

C is also known as the torsional rigidity of the material of the wire.

The total work done in twisting the wire through an angle θ is written as

$$W = \int_0^\theta C\theta d\theta = \frac{1}{2} C\theta^2 \qquad (7.22)$$

Using Eq. (7.21) in Eq. (7.22), we write

$$W = \frac{\eta\pi r^4}{4l} \theta^2 \qquad (7.23)$$

The energy used in doing this work is stored up in the wire and is known as strain energy.

7.6. BENDING OF BEAMS

Beam

A beam is defined as a rod or a bar of uniform rectangular or circular cross-section. Its length is very large as compared to its thickness or radius. A beam consists of a number of thin plane or cylindrical layers parallel or co-axial to each other. Each layer may be considered to be made of a number of parallel longitudinal fibres known as longitudinal filaments as shown in Fig. 7.7a

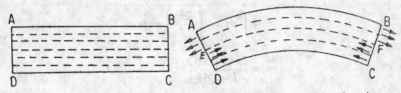

Fg.7.7a. Filaments in a beam.   Fig. 7.7b. Neutral axis *EF* in a bent beam. Filaments above *EF* extend while those below contract.

Neutral Surface

When a beam is bent its outer filments are elongated while the inner ones are shortened (Fig. 7.7b). A plane or a cylindrical layer (in a beam) which is neither elongated nor shortened during bending is called a neutral surface. In Fig. 7.7c *EFGH* is the neutral surface.

Plane of Bending

A plane (in a beam) which is normal to the neutral surface and passes through its centre is called the plane of bending. In Fig. 7.7c, *ABCD* is the plane of bending.

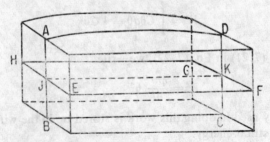

Fig. 7.7c. *JK* neutral axis. *ABCD* plane of bending. *EFGH* neutral surface.

Neutral Axis

The line of intersection of the plane of bending and the neutral surface is called the neutral axis. *JK* is the neutral axis in Fig. 7.7c.

7.7. BENDING MOMENT AND ITS DERIVATION

Fig. 7.8 shows a beam *ABCD* clamped at *AD* and loaded with a weight *W* at the end *BC*. We assume that the weight of the beam is negligible as compared to *W*. Such a beam is called a cantilever. Within the elastic limit the beam will remain in equilibrium position. Let *EBCF* be a portion of the beam cut by a transverse plane *EF*. The bending force *W* will give rise to a force of reaction *W* along *FE*. These two equal and opposite forces constitute a couple known as the bending couple. The moment of the bending couple is defined as the bending moment.

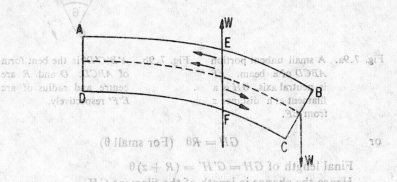

Fig. 7.8. A beam of negligible weight clamped at *AD* and loaded at *BC* (cantilever).

In equilibrium the bending couple will be balanced by an equal and opposite couple generated due to the stretching and compression of the filaments of the beam. This balancing couple is known as the restoring couple and its moment may also be referred to as the bending moment of the beam.

Fig. 7.9*a* shows a small unbent portion of a beam whose bending moment is to be determined. *EF* is its neutral axis. When *ABCD* is bent, *AB* is elongated to *A'B'* and *CD* is shortened to *C'D'*. The neutral axis *EF* = *E'F'*. Let *ABCD* be bent in the form of circular arc whose centre is situated at *O*. *R* is the radius of the arc *E'F'*; that is, *OE'* = *OF'* = *R*. Consider a filament *GH* at a distance *z* from the neutral axis. In the bent position *GH* will become *G'H'*.

Now the original length of *GH* = *EF* = *E'F'*

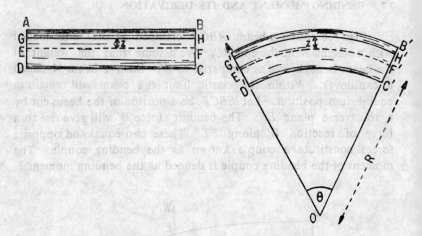

Fig. 7.9a. A small unbent portion *ABCD* of a beam. *EF* is neutral axis. *GH* is a filament at a distance *z* from *EF*.

Fig. 7.9b. *A'B'C'D'* is the bent form of *ABCD*. *O* and *R* are centre and radius of arc *E'F'* respectively.

or $$GH = R\theta \quad \text{(For small } \theta)$$

Final length of $GH = G'H' = (R + z)\,\theta$

Hence the change in length of the filament GH

$$= (R + z)\,\theta - R\theta = z\theta$$

Therefore the strain $= \dfrac{\text{change in length}}{\text{original length}}$

or $$\text{strain} = \frac{z\theta}{R\theta} = \frac{z}{R} \tag{7.24}$$

Consider a section *PQRS* of the beam at right angles to its length and plane of bending (Fig. 7.10). Obviously the stretching and the compressive forces acting on the filaments of the beam will be acting at right angles to this section. *MN* is situated on the neutral surface of the beam. Take a small area δa at a distance z from *MN*. Naturally the strain produced in a filament through this small area will be given by Eq. (7.24).

Now Young's modulus of the beam is written as

$$Y = \frac{\text{stress}}{\text{strain}} \tag{7.25}$$

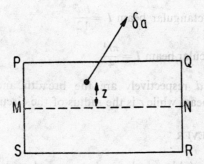

Fig. 7.10. A section $PQRS$ of the beam perpendicular to its length and plane of bending.

Using Eq. (7.24) in Eq. (7.25), we get

$$Y = \frac{\text{stress}}{z/R}$$

or

$$\text{stress} = Y\frac{z}{R}$$

Thus the force acting on area $\delta a = \frac{Yz\delta a}{R}$

The moment of this force about $MN = \frac{Y\delta az^2}{R}$

Since the moments of the forces in the upper and lower halves of section $PQRS$ are acting along the same direction, the total moment of the forces acting on the filaments of $PQRS$ is given by

$$\Sigma \frac{Y\delta az^2}{R} = \frac{Y}{R}\Sigma \delta az^2$$

or   the total moment $= \dfrac{YI}{R}$ (7.26)

where $I\ (= \Sigma\delta az^2)$ is called the geometric moment of inertia of the sectional area about MN. Eq. (7.26) represents the moment of the restoring couple of the beam. In the position of equilibrium, restoring couple is equal and opposite to the bending couple.

Hence

$$\text{bending moment} = \frac{YI}{R} \qquad (7.27)$$

The product YI is known as the flexural rigidity of the beam.

For a rectangular beam $I = \dfrac{bd^3}{12}$ $\qquad\qquad$ (7.28)

and for a circular beam $I = \dfrac{\pi r^4}{4}$ $\qquad\qquad$ (7.29)

where b and d respectively are the breadth and the depth of a rectangular beam while r is the radius of the circular beam.

7.8. CANTILEVER

A horizontal beam clamped at one end and loaded at the free end is called a cantilever, the weight of the beam being negligible as compared to the load at the free end. Fig. 7.11 shows a cantilever AB of length l clamped at A and loaded at B with a weight W. The end B is deflected to B' under the action of W. Let $BB' = \delta$.

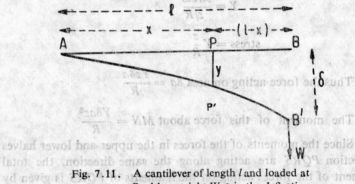

Fig. 7.11. A cantilever of length l and loaded at
B with a weight W. δ is the deflection
from mean position.

Consider a section of the cantilever at P situated at a distance x from the end A. P is deflected to P' when the beam is loaded. Let $PP' = y$.

Now the moment of the bending couple about P is given by

$$C_b = W(l - x) \qquad\qquad (7.30)$$

From Eq. (7.27), the bending moment $= \dfrac{YI}{R}$ $\qquad\qquad$ (7.31)

where Y is the Young's modulus of the cantilever, I is its geometric moment of inertia and R is the radius of curvature of the neutral axis of the section of the beam (at P) under consideration. In equilibrium position C_b is balanced by the bending moment. Hence from Eqs. (7.30) and (7.31), we have

$$\frac{YI}{R} = W(l - x) \qquad (7.32)$$

Now the radius of curvature R is given by

$$R = \frac{\left[1 + \left(\dfrac{dy}{dx}\right)\right]}{\dfrac{d^2y}{dx^2}} \qquad (7.33)$$

where y is the depression of the beam at P. As y is small $\left(\dfrac{dy}{dx}\right)^2$ can be neglected as compared to 1.

Therefore Eq. (7.33) can be written as

$$R = \frac{1}{\dfrac{d^2y}{dx^2}}$$

or

$$\frac{1}{R} = \frac{d^2y}{dx^2} \qquad (7.34)$$

Substituting Eq. (7.34) in Eq. (7.32), we write

$$YI\frac{d^2y}{dx^2} = W(l - x)$$

or

$$\frac{d^2y}{dx^2} = \frac{W}{YI}(l - x) \qquad (7.35)$$

Integrating Eq. (7.35), we get

$$\frac{dy}{dx} = \frac{W}{YI}\left(lx - \frac{x^2}{2}\right) + C_1 \qquad (7.36)$$

where C_1 is constant of integration and can be determined by suitable boundary conditions.

At $x = 0$, $y = 0$ and $\dfrac{dy}{dx} = 0$

Hence from Eq. (7.36), we have $C_1 = 0$. Thus Eq. (7.36) can be rewritten as

$$\frac{dy}{dx} = \frac{W}{YI}\left(lx - \frac{x^2}{2}\right) \qquad (7.37)$$

Integrating Eq. (7.37), we get

$$y = \frac{W}{YI}\left(\frac{lx^2}{2} - \frac{x^3}{6}\right) + C_2$$

where C_2 is constant of integration and its value is zero since $y = 0$ at $x = 0$. Therefore the depression of the beam at P is given by

$$y = \frac{W}{YI}\left(\frac{lx^2}{2} - \frac{x^3}{6}\right) \tag{7.38}$$

At $x = l$, $y = \delta$. Hence Eq. (7.38) gives

$$\delta = \frac{W}{YI}\left(\frac{l^3}{2} - \frac{l^3}{6}\right)$$

or $\qquad\qquad\qquad \delta = \frac{Wl^3}{3YI} \tag{7.39}$

Eq. (7.39) gives the deflection or the depression of the cantilever at the free end B.

For a rectangular cantilever

$$\delta = \frac{4Wl^3}{Ybd^3} \tag{7.40}$$

and for a cylindrical cantilever

$$\delta = \frac{4Wl^3}{3Y\pi r^4} \tag{7.41}$$

7.9. BEAM LOADED AT THE CENTRE AND SUPPORTED AT ITS ENDS

Consider a beam AB of length l supported at its ends by two knife-edges K_1 and K_2 and loaded at the centre O by a load W (Fig. 7.12). The weight of the beam is negligible as compared to the load W. The reaction at each knife-edge is $W/2$ and the depression at O is maximum and is equal to δ (say).

The beam AB can be imagined to be made of two inverted cantilevers each of length $l/2$ and clamped at O. Each cantilever is deflected upward at A and B by a load $W/2$.

Hence the depression δ at O can be obtained by replacing W by $W/2$ and l by $l/2$ in Eq. (7.39). Thus δ is given by

$$\delta = \frac{\left(\frac{W}{2}\right)\left(\frac{l}{2}\right)^3}{3YI}$$

or $\qquad\qquad\qquad \delta = \frac{Wl^3}{48YI} \tag{7.42}$

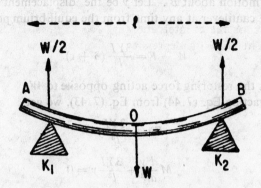

Fig. 7.12.   A beam *AB* loaded at the centre
O with *W*.   *W*/2 is the reaction
at each end.

7.10.   TRANSVERSE VIBRATIONS OF A LOADED CANTILEVER

Figure 7.13 shows a cantilever executing simple harmonic motion. When a loaded cantilever is depressed further from its equilibrium position, it starts performing simple harmonic motion about its original depressed position.   From Eq. (7.39), we have

$$\delta = \frac{Wl^3}{3YI}$$

or
$$W = \frac{3YI}{l^3}\,\delta \qquad\qquad (7.43)$$

Here $W\,(= Mg)$ is the load or the force which causes the original depression δ of the cantilever.   If the mass M of the load is slightly

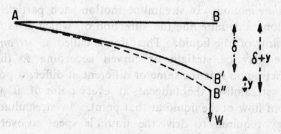

Fig. 7.13. A vibrating cantilever.

depressed further to B'' and released, the cantilever performs simple harmonic motion about B'. Let y be the displacement of the free end of the cantilever at any time from the equilibrium position (B'), then

$$W - F = \frac{3YI}{l^3}(\delta + y) \qquad (7.44)$$

where F is the restoring force acting opposite to W.

Subtracting Eq. (7.44) from Eq. (7.43), we get

$$F = -\frac{3YI}{l^3} y$$

or

$$M \frac{d^2y}{dt^2} + \frac{3YI}{l^3} y = 0$$

or

$$\frac{d^2y}{dt^2} + \frac{3YI}{Ml^3} y = 0 \qquad (7.45)$$

Here $\dfrac{d^2y}{dt^2}$ is the acceleration produced in M at the free end at that time. Eq. (7.45) represents a simple harmonic motion whose time period is given by

$$T = 2\pi \sqrt{\frac{Ml^3}{3YI}} \qquad (7.46)$$

7.11. STREAMLINE AND TURBULENT MOTION

When a liquid flows through a pipe in such a manner that the amount of liquid entering the pipe per second is the same as the amount of liquid leaving it per second, the motion of the liquid is called *steady motion*. In a steady motion each particle of the liquid, passing a point, has the same velocity and moves along the same path as that of its preceding particle. Such a motion is called orderly or *streamline motion*. In streamline motion each particle of liquid moves along a definite line the direction of which gives the direction of flow of the liquid. This line is called a *streamline*. A streamline may be straight or curved according to the lateral pressure acting on it is the same or different at different points. In a curved streamline the tangent at every point of it gives the direction of flow of the liquid at that point. In streamline motion the energy required to drive the liquid is spent to overcome the viscous force between the layers.

There is a definite velocity up to which the motion of a liquid remains streamline. This velocity is called the *critical velocity*. Beyond critical velocity the motion ceases to be streamline and becomes unsteady, disorderly and zig-zag. The motion at this stage is called *turbulent motion*. The energy required to drive the liquid during turbulent motion is dissipated in the formation of eddies and whirlpools in the liquid.

The distinction between streamline and turbulent motions can be demonstrated by taking a tube through which a liquid is allowed to flow with gradually increasing velocity. Below the critical velocity if a small jet of coloured matter is introduced into the tube, it will confine itself to a streamline. As the velocity of the liquid exceeds the critical velocity, the coloured matter takes up a zig-zag path and finally spreads out and fills the entire tube showing the existence of turbulent motion.

7.12. VISCOSITY

Consider a liquid flowing steadily over a solid horizontal surface XY (Fig. 7.14). The motion of the liquid being streamline, its various layers move parallel to the solid surface with different velocities. The layer in contact with XY is virtually at rest while the velocities of other layers increase uniformly with the increase in their distances from the solid surface.

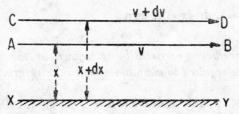

Fig. 7.14.   Steady motion of a liquid over a solid horizontal surface XY.

Consider the motion of two adjacent layers AB and CD of the liquid. AB and CD are situated at distances x and $x + dx$ and are moving with velocities v and $v + dv$ respectively. The upper layer CD moves faster and tries to increase the velocity of AB. On the other hand AB moves slower and it tends to decrease the velocity of CD. Thus the two layers mutually tend to destroy their relative motion and it appears as if some backward dragging force is acting tangentially on the layers. Hence, if the relative motion between the different layers is to be maintained, an external force must be applied to overcome this backward drag. In the absence of an external force the relative motion between the different layers will be destroyed and the liquid will cease to flow.

The property by virtue of which a liquid opposes the relative motion between its layers is called viscosity and the tangential force which tends to destroy the relative motion is known as viscous force.

As we move from the layer AB to CD (through dx) the change in velocity is dv. Thus the velocity gradient is given by dv/dx. According to Newton's law of viscous flow for a streamline motion, the viscous force F acting between two layers is given by

$$F \propto -A \frac{dv}{dx}$$

or
$$F = -\eta A \frac{dv}{dx} \tag{7.47}$$

where A is the area of each layer, dv/dx is the velocity gradient and η is a constant called the coefficient of viscosity of the liquid. The negative sign signifies that the viscous force is opposed to the flow of the liquid. Obviously the external force needed to maintain the relative motion between AB and CD must be equal and opposite to F, i.e., $-F$. Thus for an external force Eq. (7.47) reduces to

$$F = \eta A \frac{dv}{dx} \tag{7.48}$$

Let $A = 1$, $dv/dx = 1$, then
$$F = \eta \tag{7.49}$$

Hence the coefficient of viscosity is defined as the tangential force per unit area required to maintain a unit velocity gradient normal to the layer.

The C.G.S. unit of coefficient of viscosity is called poise. This is equal to dynes per square centimetre per unit velocity gradient.

Viscosity is sometimes called internal friction of the liquid. But the only common point between solid friction and viscosity is that both oppose the relative motion between the layers. The solid friction is independent of the area of contact of the solid surfaces and the relative velocity between them, but the viscous force depends on both, the area of the layers and the relative velocity between them. It also depends on the distance of separation between the layers.

7.13. CRITICAL VELOCITY AND REYNOLDS' NUMBER

The critical velocity is defined as the maximum velocity of a liquid up to which its motion remains streamline. It is found to depend on the coefficient of viscosity, the density of the liquid and the radius of the tube through which liquid is flowing.

Using the method of dimensions the critical velocity may be expressed as

$$V_c = k\frac{\eta}{\rho r} \tag{7.50}$$

where V_c is the critical velocity, η is coefficient of viscosity, ρ is density of the liquid, r is the radius of the tube and k is a constant called Reynold's number. For narrow tubes the value of k is nearly 1000. Thus the critical velocity of a liquid is

(a)   directly proportional to η

(b)   inversely proportional to ρ

and (c)   inversely proportional to r.

Hence the motion will tend to be streamline for liquids of high coefficient of viscosity and low density when they are flowing through tubes of narrow bore.

7.14. FLOW OF A LIQUID THROUGH A CAPILLARY TUBE

Poiseuille's Formula. Fig. 7.15a shows a liquid flowing through a tube AB of radius r and length l. Let p be the pressure difference between the ends of the tube. We also assume that the

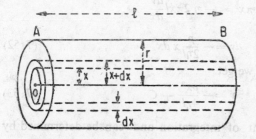
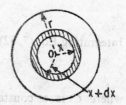

Fig. 7.15a.   A liquid flowing through a narrow cylindrical tube. Dotted lines show a coaxial cylindrical layer of radius x and thickness dx.

Fig. 7.15b.   Front section of the tube.

flow of the liquid satisfies the following conditions:

(i)   The motion of the liquid is streamline and is parallel to the axis of the tube.
(ii) The lateral pressure across the tube is constant.
(iii) The layer of the liquid in contact with the tube is at rest.
(iv) The velocity of the layer increases continuously towards the axis of the tube.

Now consider a cylindrical layer of thickness dx of the liquid coaxial with the tube. Let the inner radius of the cylindrical layer be x. Fig. 7.15b shows the front section of the tube. If v is the velocity of the liquid at a distance x from the axis of the tube, the velocity gradient across the elementary cylindrical layer is dv/dx. Applying Newton's law of viscous force

$$F = -\eta A \frac{dv}{dx}$$

or
$$F = -\eta\, 2\pi x l \frac{dv}{dx} \qquad (7.51)$$

where F is the viscous force, $A\ (=2\pi x l)$ and η are respectively the area of the cylindrical layer of the liquid and the coefficient of viscosity.

Now the external force due to the pressure difference p across the layer, is given by $p\pi x^2$. As there is no acceleration of the liquid in a steady flow, the external force must be equal to the viscous force; that is,

$$p\pi x^2 = -\eta\, 2\pi x l \frac{dv}{dx}$$

or
$$dv = -\frac{p}{2\eta l} x\, dx \qquad (7.52)$$

Integrating Eq. (7.52), we get

$$v = -\frac{px^2}{4\eta l} + C \qquad (7.53)$$

where C is a constant of integration and can be determined by initial conditions. Now at $x = r$, $v = 0$.
Hence

$$0 = -\frac{pr^2}{4\eta l} + C$$

or
$$C = \frac{pr^2}{4\eta l}$$

Substituting this value of C in Eq. (7.53), we have

$$v = \frac{pr^2}{4\eta l} - \frac{px^2}{4\eta l}$$

or
$$v = \frac{p}{4\eta l}(r^2 - x^2) \tag{7.54}$$

Eq. (7.54) gives the velocity of the layer at a distance x from the axis of the tube. The variation of v with x is shown in Fig. 7.16. Now the volume dV of the liquid flowing per second through the

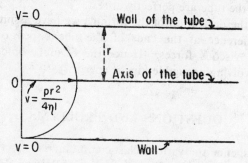

Fig. 7.16. Variation of v with x, the distance of a layer from axis of the tube. The graph is a parabola.

cyclindrical layer under consideration is given by

$$dV = \text{velocity} \times \text{the area of the layer of thickness } dx$$

or
$$dV = v\, 2\pi x\, dx \tag{7.55}$$

Using Eq. (7.54) in Eq. (7.55), we get

$$dV = \frac{p}{4\eta l}(r^2 - x^2)\, 2\pi x\, dx \tag{7.56}$$

The total volume V of the liquid flowing through the tube per second is obtained by integrating Eq. (7.56) between the limits $x = 0$ and $x = r$. Thus

$$V = \int_0^r \frac{p}{4\eta l}(r^2 - x^2)\, 2\pi x\, dx$$

or
$$V = \frac{p\pi}{2\eta l} \int\limits_{0}^{r} (r^2 - x^2)\, x\, dx$$

or
$$V = \frac{p\pi}{2\eta l} \left[\frac{r^2 x^2}{2} - \frac{x^4}{4} \right]_0^r$$

or
$$V = \frac{p\pi r^4}{8\eta l} \qquad\qquad (7.57)$$

Eq. (7.57) is known as *Poiseuille's formula*. This formula has the following limitations:

    (i) It holds good only for streamline motion.

    (ii) It is true only when the layers of the liquid in contact with the walls of the tube are perfectly at rest.

    (iii) In the derivation of the formula we have assumed that the pressure difference at the ends of the tube is spent only in overcoming the viscous forces. Hence the formula holds for fluid velocities involving negligible kinetic energy.

QUESTIONS AND PROBLEMS

7.1. (a) Define the various elastic constants.

      (b) Show that the limiting values of Poisson's ratio are -1 and 0.5.

7.2. Define longitudinal, volume and shearing stresses. Show that a shearing stress is equivalent to an equal linear tensile stress and an equal compression stress at right angles to each other.

7.3. Define Young's modulus, Bulk modulus and Poisson's ratio. Prove the relation

$$K = \frac{Y}{3(1 - 2\sigma)}$$

where the symbols have their usual meanings.

7.4. State and discuss Hooke's law. Show that the Young's modulus Y, modulus of rigidity η and Poisson's ratio σ are related by the equation

$$Y = 2\eta(1 + \sigma)$$

7.5. What do you understand by the angle of twist and the angle of shear? Deduce an expression for the couple required to twist a cylindrical rod or a wire.

7.6. (a) Define the following terms: (a) Beam, (b) Neutral surface, (c) Neutral axis, (d) Bending moment.

(b) Derive an expression for the bending moment for a beam.

7.7. What is a cantilever? Derive an expression for the depression produced when a cantilever is loaded at its free end.

7.8. Discuss the theory of bending of a beam loaded at the centre and supported at its ends. How can this be used to determine the Young's modulus of a uniform rectangular rod?

7.9. Discuss the theory of the transverse vibrations of a loaded cantilever.

7.10. (a) What do you understand by streamline and turbulent motions?

(b) Define the following quantities:
(i) Viscosity, (ii) Coefficient of viscosity, (iii) Critical velocity, (iv) Reynold's number.

7.11. Discuss the origin of a viscous force. Derive Poiseuille's formula for the rate of flow of a liquid through a capillary tube. What are the limitations of the formula?

7.12. A wire 300 cm long and 0.625 sq. cm in cross-section is found to stretch by 0.3 cm under a tension of 1200 kg. Determine the Young's modulus of the wire.

$$(18.8 \times 10^{11} \text{ dynes/cm}^2)$$

7.13. A wire 4 metres long 0.30 mm diameter is stretched by a force of 800 gm weight. If the extension in length amounts to 1.5 mm, calculate the energy stored in the wire.

$$(5.88 \times 10^4 \text{ ergs})$$

7.14. A plate of metal 100 sq. cm in area rests on a layer of castor oil 2 mm thick whose $\eta = 15.5$ poise. Calculate the horizontal force required to move the plate with a speed of 3 cm/sec.

$$(23, 250 \text{ dynes})$$

7.15. Calculate the mass of water flowing in 10 minutes through a tube of 0.1 cm in diameter, 40 cm long, if there is a constant pressure head of 20 cm of water. η for water is 0.0082 C.G.S. units.

$$(88.03 \text{ gm})$$

Part Three

STATISTICAL PHYSICS

8. Classical Maxwell-Boltzmann Statistics

8.1. INTRODUCTION

A system consisting of a large number of particles can be described by either microscopic or macroscopic approach. In the former the motion of the individual particles comprising the system is considered by using the familiar laws of classical mechanics, viz., the Newton's laws of motion. In macroscopic approach the attention is focussed mainly on the system as a whole without giving much consideration to the individual particles.

Statistical physics is basically the macroscopic approach applied to a system of particles by using the laws of probability or chance. This enables us to obtain some sort of average values relating to the system. These average values correspond to the bulk properties of the system. In the case of a gas, for instance, the bulk properties are pressure, temperature etc. Having obtained the bulk properties or the average values for the system, the values for the individual particles are obtained. The macroscopic properties of the system are thus related to the microscopic properties in the statistical description of the system. In statistical mechanics the position or velocity (or momentum) of the individual particles of the system at any particular instant of time can be determined. But this does not tell us about the past history of the particle, namely, the position or the momentum of a particle before a particular instant of time.

8.2. PHASE SPACE

The state of a system of particles is completely specified by knowing the positions and the momenta of each and every particle of the system. The position of a particle in the ordinary three-dimensional space can be specified by the coordinates x, y and z. In the same way the momentum of a particle can be specified in a three-dimensional momentum space by the coordinates p_x, p_y and p_z. Generalising this conception, it is possible to specify the position and the momentum of a particle in the combined position and momentum space by the six coordinates x, y, z, p_x, p_y and p_z. This six-dimensional space is called *phase space*. A point in phase space corresponds to a particular position and momentum. Thus every particle of a system can be completely specified by a point in phase space.

The position of a particle in the three-dimensional space alone can be specified without any uncertainty. The momentum of a particle too can be accurately specified in the momentum space. But when we consider phase space it is not possible to simultaneously specify both the position and momentum of a particle with exactness. This is a consequence of the uncertainty principle due to Hiesenberg according to which an uncertainty of the order of h (Planck's constant) exists in the simultaneous measurement of the position and momentum of a particle. Stated mathematically

$$dx\, dp_x \geqslant h$$

$$dy\, dp_y \geqslant h$$

$$dz\, dp_z \geqslant h \qquad (8.1)$$

A volume element in the ordinary space is defined as $V_s = dx\, dy\, dz$. In the same manner a volume element in the momentum space can also be defined as $V_m = dp_x\, dp_y\, dp_z$. Thus we can define a volume element in the phase space (called phase volume) as

$$V_p = V_s \times V_m$$

This can be written as

$$V_p = dx\, dy\, dz\, dp_x\, dp_y\, dp_z$$

Using Eq. (8.1) this leads to

$$V_p \geqslant h^3$$

Thus the particle can be found anywhere within a six-dimensional cell whose minimum volume is of the order of h^3. The particle will therefore not be exactly located about the point (x, y, z, p_x, p_y, p_z) in the phase space but can only be found about this point within a cell of volume h^3.

8.3. LAWS OF PROBABILITY AND CHANCE

If an event can occur in a large number of different ways, then the probability that the event will occur in a particular way is defined as

$$\text{Probability} = \frac{\text{Number of favourable occurrences}}{\text{Total number of occurrences}}$$

Let us consider the case of two distinguishable coins, say one of copper and another of silver. If the coins be tossed a large number of times, the various possible events are:

(a) Heads of both coins uppermost
(b) Tails of both coins uppermost
(c) Head of one and tail of the other uppermost
(d) Tail of one and head of the other uppermost

If heads and tails be represented by a and b respectively and the two coins are designated by 1 and 2 then the possible occurrences are a_1a_2, b_1b_2, a_1b_2 and b_1a_2. Probability of occurrence of any of the four events is the same, i.e., equal to 1/4. Consider the event a_1a_2. The probability of getting the head of the coin '1' uppermost is 1/2; this is also the probability of getting the head of the coin '2' uppermost provided the two coins are tossed independently. However, if the two coins are tossed simultaneously the probability of getting the heads of both the coins uppermost is 1/4 or $\frac{1}{2} \times \frac{1}{2}$, i.e., the product of the probabilities in the independent tosses of the coins. Thus *the probability of a composite event is the product of the probabilities of the individual and independent component events.*

The above four events can also be obtained as terms in the product of the factors $a_1 + b_1$ and $a_2 + b_2$,

$$(a_1 + b_1)(a_2 + b_2) = a_1a_2 + a_1b_2 + b_1a_2 + b_1b_2$$

Now, instead of two distinguishable coins as in the above example, we consider tossing of two coins which are indistinguishable from each other, say two silver coins. For this case there are only three distinguishable events, viz., a^2, ab and b^2. These are given by the terms in the binomial expansion of $(a + b)^2$,

$$(a + b)^2 = a^2 + 2ab + b^2$$

In this case the probability of getting either two heads uppermost (a^2) or two tails uppermost (b^2) is 1/4. But the probability of getting head of one and tail of the other uppermost (ab) is $2 \times 1/4$ or 1/2. Thus we see that the probability of any event is proportional to the coefficient of the term representing that particular event. In general, the probability of any event can be obtained as the ratio of the coefficient of the term representing the event to the sum of the coefficients of all the terms (i.e., the total number of events).

Case of n Indistinguishable Coins

Let's now extend our earlier considerations to the case of n identical coins indistinguishable from one another. If they are tossed a large number of times, the possible events or complexions will be given by the terms in the binomial expansion of $(a + b)^n$

$$(a + b)^n = a^n + {}^nC_1 a^{n-1}b + {}^nC_2 a^{n-2}b^2 + \ldots + {}^nC_r a^{n-r}b^r + \ldots + b^n$$

The term $a^{n-r}b^r$ in the above expansion signifies that $n - r$ coins have heads uppermost while r coins have tails uppermost. The probability for the occurrence of the complexion $a^{n-r}b^r$ is clearly

$$W = \frac{{}^nC_r}{{}^nC_0 + {}^nC_1 + {}^nC_2 + \ldots + {}^nC_n}$$

or

$$W = \frac{n!}{r! \, s!} \times \frac{1}{\sum\limits_{r=0}^{n} {}^nC_r} \tag{8.2}$$

where $s = n - r$.

Eq. (8.2) finally simplifies to

$$W = \frac{n!}{r! \, s!} \times 2^{-n} \tag{8.3}$$

since

$$\sum_{r=0}^{n} {}^nC_r = (1 + 1)^n = 2^n$$

In the binomial expansion of $(a + b)^n$, the sum of the coefficients of the various terms is 2^n. This represents the total number of possible events or complexions. Therefore, the a priori or the intrinsic probability of any complexion is $1/2^n$ or 2^{-n}.

Complexion Having Maximum Probability

Of the $(n + 1)$ different or distinguishable complexions, the complexion $a^s b^r$ will possess maximum probability if ^nC_r is maximum. From our knowledge of elementary algebra we know that ^nC_r is maximum when $r = \dfrac{n}{2}$. The maximum probability value from Eq. (8.3) is given by

$$W_{max} = \frac{n!}{\left(\dfrac{n}{2}!\right)^2} 2^{-n} \tag{8.4}$$

Case of n Dices

Now instead of coins we toss n dices. Each dice has six faces which we denote by the letters a, b, c, d, e and f. Let the dices be thrown a large number of times. Consider the complexion

$$a^{n_1} b^{n_2} c^{n_3} d^{n_4} e^{n_5} f^{n_6}$$

which means that n_1 dices fall with face a uppermost, n_2 dices fall with face b uppermost, and so on. In a manner similar to the one for the case of coins, it can be shown that the probability for the occurrence of the complexion $a^{n_1} b^{n_2} \ldots f^{n_6}$ is given by

$$W = \frac{n!}{n_1! \, n_2! \, n_3! \, n_4! \, n_5! \, n_6!} \times 6^{-n}$$

where 6^{-n} represents the a priori probability for each complexion.

General Case

We now pass on to a more general case by considering a board having p cells arranged side-by-side. Let a large number of n balls be thrown into the cells from a distance. We denote the cells by $a_1, a_2, a_3, \ldots, a_p$. Let n_1 balls be found in the cell a_1, n_2 balls in the cell a_2, and so on. Such a complexion is represented by $a_1^{n_1} a_2^{n_2} \ldots a_p^{n_p}$. The mathematical probability for this complexion is

$$W = \frac{n!}{n_1! \, n_2! \, n_3! \ldots n_p!} p^{-n} \tag{8.5}$$

Note that for this case the a priori probability for each complexion is p^{-n}, the a priori probability for each cell being $1/p$.

We rewrite Eq. (8.5) in the form

$$W = \frac{n!}{\pi n_r!} p^{-n} \tag{8.6}$$

where $\pi n_r! = n_1! \, n_2! \, n_3! \, \ldots .$

When n is large, Eq. (8.6) can be simplified by applying a well-known theorem called Stirling's theorem. According to this theorem, for large n

$$\log n! = n \log n - n \tag{8.7}$$

Using Stirling's theorem Eq. (8.6) simplifies to

$$\log W = n \log n - n - \log \pi n_r! - n \log p$$

or $\log \{ W \pi n_r! \} = n \log n - (1 + \log p) n$

8.4. THE PROBABILITY OF A DISTRIBUTION

In the preceding example we considered the throwing of n balls into p cells. We throw the balls into the cells and note how the balls arrange themselves in the various cells. If we repeat this experiment a large number of times it will be found that a certain distribution of the balls among the various cells occurs more often than any other distribution. This is called the most probable distribution, i.e., the distribution which is most likely to be found. Extending this analogy to a system consisting of n particles, the most probable distribution of the system can be obtained by investigating how the particles of the system distribute themselves in phase space. Of the various possible distributions of particle positions and momenta, the one having the maximum probability can be easily selected by calculating the probabilities of all the distributions permitted by the nature of the system. We can then assert that the system tends to behave according to that particular (most probable) distribution of particles in phase space.

8.5. CLASSICAL MAXWELL-BOLTZMANN STATISTICS

We now pass on to examine how a fixed total amount of energy is distributed among the various members of a system comprising n particles. It is assumed that the particles are identical but are sufficiently widely separated to be distinguished. The molecules of a gas provide an example of particles of this kind. Let these particles be distributed among the s cells of the system. The cells are designated as A_1, A_2, A_3, ..., A_s and can accommodate n_1, n_2, n_3, ..., n_s particles respectively. The n_1 particles in the cell A_1 have energy E_1, n_2 particles in the cell A_2 have energy E_2, and so on. Our problem is to find out how the total energy is distributed among the various particles of the system, that is, how many particles (on the average) have energy E_1, how many have energy E_2, and so on. We shall now apply the methods of statistical mechanics to calculate this energy distribution.

The a priori probability g_i that a particle will occupy the ith cell is proportional to the volume of that cell. This is because the cell with larger volume has a greater probability for the occupation of a particle than a cell with a smaller volume. Therefore g_i is given by

$$g_i = \frac{v_i}{V}$$

where v_i is the volume of the ith cell and $V = \Sigma\, v_i$ is the total volume of all the cells, i.e., of the entire system.

We now know that the a priori probability that a particle will occupy the ith cell is g_i. Therefore, applying the probability law the a priori probability that two particles will occupy the ith cell is $g_i \times g_i$, i.e., g_i^2. Similarly, the a priori probability that n_i particles will occupy the ith cell will be $g_i^{n_i}$. So, $g_1^{n_1}$, $g_2^{n_2}$, $g_3^{n_3}$, ..., $g_s^{n_s}$ are the respective a priori probabilities of the various cells. Thus the a priori probability of any particular distribution of n particles among the s cells is obtained by taking the products of all the a priori probabilities of the form $g_i^{n_i}$, namely,

$$g_1^{n_1}\, g_2^{n_2}\, g_3^{n_3}\, \ldots\, g_s^{n_s}$$

The mathematical probability for the distribution of n particles among the s cells is written in analogy with Eq. (8.5) in the form

$$W = \frac{n!}{n_1!\, n_2!\, n_3!\, \ldots\, n_s!}\, g_1^{n_1}\, g_2^{n_2} \ldots g_s^{n_s} \tag{8.8}$$

When the cells are all of the same size, the a priori probabilities g_1, g_2, ..., g_s are each equal to $1/s$. For this case Eq. (8.8) reduces to

$$W = \frac{n!}{n_1!\, n_2!\, n_3!\, \ldots\, n_s!}\, s^{-n}$$

which is of the same form as Eq. (8.5).

We rewrite Eq. (8.8) in the form

$$W = \frac{n!}{\pi n_s!} \, \pi g_s^{n_s} \tag{8.9}$$

where

$$\begin{aligned} \pi n_s! &= n_1! \, n_2! \ldots n_s! \\ \pi g_s^{n_s} &= g_1^{n_1} \, g_2^{n_2} \ldots g_s^{n_s} \end{aligned} \tag{8.10}$$

Taking the logarithm of both sides of Eq. (8.9)

$$\log W = \log n! + \log(\pi g_s^{n_s}) - \log(\pi n_s!) \tag{8.11}$$

Using Stirling's theorem (Eq. (8.7)) for large values of n, the various factors in Eq. (8.11) simplify as follows:

$$\log n! = n \log n - n$$

$$\begin{aligned} \log(\pi g_s^{n_s}) &= \log g_1^{n_1} + \log g_2^{n_2} + \ldots + \log g_s^{n_s} \\ &= n_1 \log g_1 + n_2 \log g_2 + \ldots + n_s \log g_s \\ &= \Sigma \, n_s \log g_s \end{aligned}$$

$$\begin{aligned} \log(\pi n_s!) &= \log(n_1! \, n_2! \, n_3! \, \ldots \, n_s!) \\ &= \log n_1! + \log n_2! + \ldots + \log n_s! \\ &= n_1 \log n_1 - n_1 + n_2 \log n_2 - n_2 + \ldots + n_s \log n_s - n_s \\ &= (n_1 \log n_1 + n_2 \log n_2 + \ldots + n_s \log n_s) \\ &\qquad\qquad\qquad\qquad\qquad - (n_1 + n_2 + \ldots + n_s) \\ &= \Sigma \, n_s \log n_s - \Sigma \, n_s \\ &= \Sigma n_s \log n_s - n \end{aligned}$$

Substitution of these values in Eq. (8.11) yields

$$\log W = n \log n + \Sigma \, n_s \log g_s - \Sigma \, n_s \log n_s \tag{8.12}$$

For obtaining the distribution law corresponding to the equilibrium state of the system, the essential condition is that the system must possess maximum probability corresponding to the most probable distribution of particles in phase space. If the probability is a maximum, its logarithm also is a maximum. So the condition of maximum probability is

$$\delta(\log W) = 0 \tag{8.13}$$

Besides the condition stated by Eq. (8.13) there are two subsidiary conditions to be fulfilled, namely, that the total number of particles and the total energy of an isolated system must be constant. These conditions are expressed by the following equations

$$\delta \Sigma n_s = 0 \tag{8.14}$$

$$\delta \Sigma n_s E_s = 0 \tag{8.15}$$

Differentiating Eq. (8.12)

$$\delta (\log W) = 0 + \Sigma \log g_s \, \delta n_s - \Sigma (\delta n_s + \log n_s \, \delta n_s)$$

or $\quad \delta (\log W) = \Sigma \log g_s \, \delta n_s - \Sigma (1 + \log n_s) \delta n_s \tag{8.16}$

When Eq. (8.13) is incorporated in Eq. (8.16) we get

$$\Sigma (\log g_s - \log n_s) \, \delta n_s - \Sigma \, \delta n_s = 0$$

or $\quad \Sigma \log \left(\dfrac{g_s}{n_s}\right) \delta n_s = 0 \tag{8.17}$

since from Eq. (8.14)

$$\Sigma \, \delta n_s = 0$$

Eqs. (8.14) and (8.15) may be rewritten as

$$\Sigma \, \delta n_s = 0 \tag{8.18}$$

$$\Sigma \, E_s \delta n_s = 0 \tag{8.19}$$

Eqs. (8.17), (8.18) and (8.19) may be combined by the Lagrange's method of undetermined multipliers. We multiply Eq. (8.18) by $-\alpha$ and Eq. (8.19) by $-\beta$ and then add the resulting equations to Eq. (8.17). The result is

$$\Sigma \left[\log \left(\dfrac{g_s}{n_s}\right) - \alpha - \beta E_s \right] \delta n_s = 0 \tag{8.20}$$

In Eq. (8.20) α and β are undetermined multipliers and are called Lagrange's multipliers. For Eq. (8.20) to hold good for all values of s, the quantity in brackets must vanish, i.e.,

$$\log \left(\dfrac{g_s}{n_s}\right) - \alpha - \beta E_s = 0$$

or $\quad \dfrac{g_s}{n_s} = \exp (\alpha + \beta E_s)$

whence $\quad n_s = \dfrac{g_s}{f} \exp (-\beta E_s) \tag{8.21}$

where $\quad f = \exp (\alpha)$

Eq. (8.21) is called Maxwell-Boltzmann distribution law; β and f are called the distribution modulus and the degeneracy parameter respectively. Eq. (8.21) gives the number of particles n_s possessing energy E_s. As g_s depends on the size of the cells, it follows from Eq. (8.21) that among cells of equal size those having lower energies will be more filled or occupied than cells having higher energies. Thus the number of particles decreases exponentially with increasing energy, or that n_s is an exponentially decreasing function of E_s.

8.6. MAXWELL DISTRIBUTION LAW OF MOLECULAR VELOCITIES

In the derivation of Maxwell-Boltzmann distribution law we assumed the particles to be identical but distinguishable. This assumption is perfectly valid for the molecules of a gas. Thus the distribution law given by Eq. (8.21) can be used to study the distribution of velocities among the various molecules of a gas. To do so, our first task is to evaluate the values of the constants f and β appropriate to the above case. It is convenient to assume a continuous distribution of molecular energies rather than a discrete set of energy values $E_1, E_2, E_3, \ldots$ etc. Thus energy quantization can be dispensed with, because the number of molecules in a gas is usually very large.

The number of molecules with energies lying between E_s and $E_s + dE_s$ is from Eq. (8.21),

$$n_s dE_s = \frac{g_s}{f} \exp(-\beta E_s)\, dE_s \qquad (8.22)$$

In terms of molecular momenta since $E = p^2/2m$, we have

$$n_s dp = \frac{g_s}{f} \exp(-\beta p^2/2m)\, dp \qquad (8.23)$$

We know that g_s is the a priori probability that a molecule will possess the energy E_s or momentum p. Thus $g_s dp$ may be interpreted as the a priori probability that the molecule will have momentum between p and $p + dp$. So $g_s dp$ can be obtained by calculating the number of cells in phase space in the specified momentum range within which such a molecule may exist. If τ be the volume of a cell and V_p the total phase volume in the momentum range between p and $p + dp$, then the total number of cells in the momentum range p and $p + dp$ is V_p/τ. This must be equal to

$g_s dp$. The phase-space volume V_p occupied by the molecules with specified momenta is

$$V_p = \iiint dx\, dy\, dz \iiint dp_x\, dp_y^c dp_z$$

So, we have

$$g_s dp = \frac{\iiint dx\, dy\, dz \iiint dp_x\, dp_y\, dp_z}{\tau} \qquad (8.24)$$

Strictly speaking τ in Eq. (8.24) should be equal to h^3 (see introduction to this chapter), but as we have not incorporated the concept of quantization the volume of a cell is denoted by τ only.

In Eq. (8.24) $\iiint dx\, dy\, dz$ is equivalent to the volume occupied by the gas in the ordinary position space. If this volume be denoted by V, we have

$$g_s dp = \frac{V}{\tau} \iiint dp_x\, dp_y\, dp_z \qquad (8.25)$$

In Eq. (8.25) $\iiint dp_x\, dp_y\, dp_z$ gives the volume in the momentum space between momenta p and $p + dp$. This is equal to the volume of a spherical shell having radius p and thickness dp (Fig. 8.1), i.e.

$$\iiint dp_x\, dp_y\, dp_z = 4\pi p^2\, dp \qquad (8.26)$$

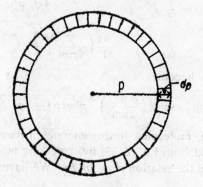

Fig. 8.1.

Substituting in Eq. (8.25)

$$g_s dp = \frac{4\pi V}{\tau} p^2 dp \qquad (8.27)$$

Hence from Eq. (8.23)

$$n_s dp = \frac{4\pi V p^2 \exp\left(-\beta p^2/2m\right)}{\tau f} dp \qquad (8.28)$$

Since in the gas we have molecules with all possible momenta between 0 and ∞, the total number of molecules in the gas (N) is obtained by integrating Eq. (8.28) between the limits 0 and ∞. Hence

$$N = \int_0^\infty n_s dp$$

which on using Eq. (8.28) gives

$$N = \frac{4\pi V}{\tau f} \int_0^\infty p^2 \exp\left(-\beta p^2/2m\right) dp \qquad (8.29)$$

Integrating

$$N = \frac{4\pi V}{\tau f} \frac{1}{4} \sqrt{\frac{2^3 \pi m^3}{\beta^3}}$$

or

$$N = \frac{V}{\tau f} \left(\frac{2\pi m}{\beta}\right)^{3/2} \qquad (8.30)$$

where we have made use of the definite integral

$$\int_0^\infty x^2 \exp\left(-ax^2\right) dx = \frac{1}{4} \sqrt{\frac{\pi}{a^3}} \qquad (8.31)$$

Hence we get

$$\frac{1}{f} = \frac{N\tau}{V} \left(\frac{\beta}{2\pi m}\right)^{3/2} \qquad (8.32)$$

Eq. (8.28) now becomes

$$n_s dp = 4\pi N \left(\frac{\beta}{2\pi m}\right)^{3/2} p^2 \exp\left(-\beta p^2/2m\right) dp \qquad (8.33)$$

The number of molecules having energies between E and $E + dE$ can be obtained from Eq. (8.33) by expressing momentum in terms of energy using the relation $p^2 = 2m E$. We have

$$dp = \frac{m}{\sqrt{2mE}} dE$$

Hence

$$n_s dE = 2\pi N \left(\frac{\beta}{\pi}\right)^{3,2} \sqrt{E} \exp\left(-\beta E\right) dE \qquad (8.34)$$

Our next step is to evaluate the parameter β. For this we obtain the total energy E_T of the assembly of molecules which is

$$E_T = \int_0^\infty E n_s \, dE \qquad (8.35)$$

Using Eq. (8.34) in Eq. (8.35) we have

$$E_T = \frac{2N}{\sqrt{\pi}} \beta^{3/2} \int_0^\infty E^{3/2} \exp(-\beta E) \, dE \qquad (8.36)$$

Eq. (8.36) yields on integration

$$E_T = \frac{3}{2} \frac{N}{\beta} \qquad (8.37)$$

where use has been made of the definite integral

$$\int_0^\infty x^{3/2} \exp(-ax) \, dx = \frac{3}{4a^2} \sqrt{\frac{\pi}{a}} \qquad (8.38)$$

According to the kinetic theory of gases, the total energy of an ideal gas (that is what we have been considering) consisting of N molecules at absolute temperature T is

$$E = (3/2) N k T \qquad (8.39)$$

Eqs. (8.37) and (8.39) agree if

$$\beta = \frac{1}{kT} \qquad (8.40)$$

Since we have now evaluated the parameters f and β, we can write the Maxwell-Boltzmann distribution law in its various forms. Thus from Eqs. (8.33) and (8.34) we have by setting $\beta = 1/kT$:

$$n_s \, dp = \frac{4N}{\sqrt{\pi}} \left(\frac{1}{2mkT}\right)^{3/2} p^2 \exp(-p^2/2mkT) \, dp \qquad (8\,41)$$

$$n_s \, dE = 2\pi N \left(\frac{1}{\pi kT}\right)^{3/2} \sqrt{E} \exp(-E/kT) \, dE \qquad (8.42)$$

Eqs. (8.41) and (8.42) respectively give the distribution laws for the momenta and energies of the molecules of a gas. The distribution law for velocities can be obtained from Eq. (8.41) by setting

$p = mv$. We have

$$n_s\, dv = 4\pi N \left(\frac{m}{2\pi kT}\right)^{3/2} v^2 \exp\left(-\frac{mv^2}{2kT}\right) dv \qquad (8.43)$$

Eq. (8.43) may be rewritten in the form

$$n_s\, dv = 4\pi N a^3 \exp\left(-bv^2\right) v^2\, dv \qquad (8.44)$$

where $\qquad b = \dfrac{m}{2kT}\;, a = \sqrt{\dfrac{b}{\pi}} \qquad (8.45)$

Eq. (8.44) was first obtained by Maxwell in 1859. This relation is therefore named after him as the Maxwell's law of distribution of molecular velocities. Eq. (8.44) is plotted in Fig. 8.2. The

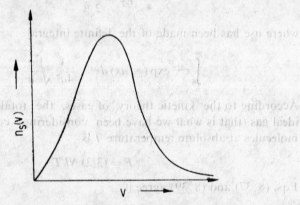

Fig. 8.2. Maxwell's velocity distribution curve.

corresponding energy distribution as given by Eq. (8.42) is plotted in Fig. 8.3. From Eq. (8.44) we see that for very small values of v the exponential term is near unity and n_s is proportional to v^2. For very large values of v the exponential term is the more important one and n_s is proportional to $\exp\left(-mv^2/2kT\right)$. Thus the distribution curve for velocity shown in Fig. 8.2 first rises parabolically and then, after attaining a maximum, falls exponentially to zero. The maximum corresponds to the most probable speed denoted by v_m.

8.7. EXPERIMENTAL VERIFICATION OF MAXWELL VELOCITY DISTRIBUTION

Direct experimental verifications of Maxwell velocity distri-

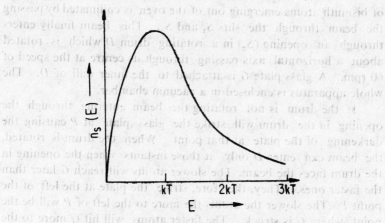

Fig. 8.3. Maxwell's energy distribution curve.

bution have been made through a number of experiments. We shall describe below two such experiments.

(i) *Zartman and Ko Experiment*

The experiment performed by I.F. Zartman and C.C. Ko in 1930–1934 is shown in Fig. 8.4. The apparatus consists of an oven O which contains bismuth vapour at about 800°C. The beam

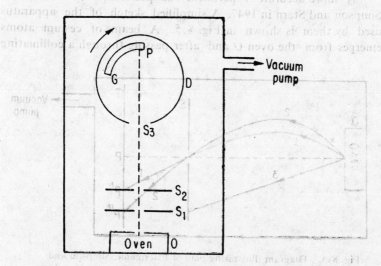

Fig. 8.4. Experimental arrangement of Zartman and Ko.

of bismuth atoms emerging out of the oven is collimated by passing the beam through the slits S_1 and S_2. This beam finally enters through an opening (S_3) in a rotating drum D which is rotated about a horizontal axis passing through its centre at the speed of 60 rpm. A glass plate G is attached to the inner wall of D. The whole apparatus is enclosed in a vacuum chamber.

If the drum is not rotating the beam entering through the opening in the drum will strike the glass plate at P causing the darkening of the plate at that point. When the drum is rotated, the beam can enter D only at those instants when the opening in the drum faces the beam. The slower atoms will reach G later than the faster ones. They, therefore, strike the plate at the left of the point P. The slower the atoms the more to the left of P will be the point where G is struck. The faster atoms will hit G more to the right. Thus the intensity of deposit on the plate depends on the velocity of the bismuth atoms. When sufficient deposit has been obtained on the plate it can be examined with the help of a spectrophotometer. The density of deposit is a measure of the speed distribution of the bismuth atoms; and this distribution agrees with the distribution predicted by Eq. (8.44).

(ii) Estermann, Simpson and Stern Experiment

A more accurate experiment was performed by Estermann, Simpson and Stern in 1947. A simplified sketch of the apparatus used by them is shown in Fig. 8.5. A beam of cesium atoms emerges from the oven O and after passing through a collimating

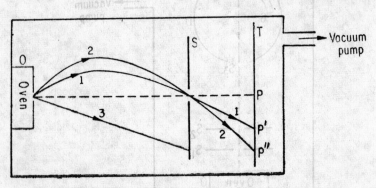

Fig. 8.5. Diagram illustrating plan of Estermann, Simpson and Stern experiment.

slit S impinges on a hot tungsten wire T. The wire is surrounded by a negatively charged cylinder which acts as the detector; this is however not shown in the diagram. The apparatus is enclosed in a vacuum chamber inside which the pressure is of the order of 10^{-8} mm of mercury. Both the slit and the tungsten wire are kept horizontal. The cesium atoms striking the tungsten wire get ionized, re-evaporate and are collected by the detector surrounding the wire thereby producing an ion current. The strength of the ion current is a measure of the number of cesium atoms striking the tungsten wire at a given point.

In the absence of a gravitational field, only those atoms which emerge in a horizontal direction will be able to pass through the slit S. They will strike the tungsten wire at the point P. However, since a gravitational field is acting the path of the various atoms will be a parabola. Atoms with greater speeds will stay in the influence of the gravitational field for lesser intervals of time as compared to the slow moving atoms. Consequently, a fast moving atom moving along the path 1 will suffer less deflection PP' as compared to the deflection (PP'') suffered by a relatively slow velocity atom moving along the path 2. An atom emerging with very small velocity is, however, pulled down by the gravitational field and thus moves along the path 3 unable to pass through S.

The detector is moved vertically downward to various positions to receive ions of different velocities. The vertical height S of the detector below the position P is a measure of the velocity of cesium atoms. A graph is plotted with the vertical height S as abscissa and the strength of the ion current (which is a measure of the number of cesium atoms) as the ordinate. The curve obtained is found to conform fairly well with the Maxwell's speed distribution.

8.8. THE LAW OF EQUIPARTITION OF ENERGY

We now deduce the law of equipartition according to which the total energy of a dynamical system in equilibrium is equally shared among its various degrees of freedom. The degrees of freedom are *the total number of independent parameters that must be specified in order to determine the energy of a molecule.* For example, the translational kinetic energy of a molecule depending on the three velocity components of its centre of mass has three translational degrees of freedom. This translational kinetic energy is equally distributed among all the three degrees of freedom. In other words, we have equipartition of energy among the three translational degrees of freedom. The equipartition law was first arrived at by Maxwell in 1859 and was later shown by Boltzmann to apply to the energies of rotation as well as of vibration.

The fact that the molecules are not mere geometric points but have finite size suggests that besides mass the molecules must also possess moment of inertia. Therefore, we would expect them to rotate because of the random collisions with other molecules in the gas. Furthermore, as the molecules are not perfectly rigid structures they can be expected to oscillate or vibrate as a result of collisions with other molecules. Thus the rotations and the vibrations of the molecules may give rise to degrees of freedom other than those of translational.

We now show how the energy equipartition follows from Maxwell-Boltzmann statistics. We start with the velocity distribution given by Eq. (8.44) and rewrite it in the form

$$dN = Na^3 \exp\left[-b(v_x{}^2 + v_y{}^2 + v_z{}^2)\right] dv_x\, dv_y\, dv_z \qquad (8.46)$$

Note that Eq. (8.46) gives the number of gas molecules with velocities lying between v_x and $v_x + dv_x$; v_y and $v_y + dv_y$; and v_z and $v_z + dv_z$.

The average kinetic energy per molecule is given by

$$\bar{E} = \frac{1}{N} \int E\, dN \qquad (8.47)$$

where E is the kinetic energy possessed by each of the dN molecules,

$$E = \tfrac{1}{2}m\,(v_x{}^2 + v_y{}^2 + v_z{}^2) \qquad (8.48)$$

The average kinetic energy is therefore

$$\bar{E} = \frac{1}{N}\left[\int \tfrac{1}{2}mv_x{}^2\, dN + \int \tfrac{1}{2}mv_y{}^2\, dN + \int \tfrac{1}{2}mv_z{}^2\, dN\right]$$

$$= I_x + I_y + I_z \qquad (8.49)$$

The integrals I_x, I_y and I_z are defined by

$$I_x = \frac{1}{N} \int \tfrac{1}{2}mv_x{}^2\, dN$$

$$I_y = \frac{1}{N} \int \tfrac{1}{2}mv_y{}^2\, dN \qquad (8.50)$$

$$I_z = \frac{1}{N} \int \frac{1}{2} m v_z^2 \, dN$$

The above integrals can be easily evaluated. Using Eq. (8.46) I_x simplifies to

$$I_x = \frac{1}{2} m \frac{\iiint v_x^2 \exp[-b(v_x^2 + v_y^2 + v_z^2)] \, dv_x \, dv_y \, dv_z}{\iiint \exp[-b(v_x^2 + v_y^2 + v_z^2)] \, dv_x \, dv_y \, dv_z} \qquad (8.51)$$

The integrations are to be carried out between the limits 0 and ∞. As the velocity components v_x, v_y and v_z of a molecule are all independent, each of the integrals in the numerator and the denominator of Eq. (8.51) may be regarded as a product of two integrals. Hence

$$I_x = \frac{1}{2} m \frac{\int v_x^2 \exp(-b v_x^2) \, dv_x \iint \exp[-b(v_y^2 + v_z^2)] \, dv_y \, dv_z}{\int \exp(-b v_x^2) \, dv_x \iint \exp[-b(v_y^2 + v_z^2)] \, dv_y \, dv_z} \qquad (8.52)$$

On evaluation this gives

$$I_x = \frac{1}{2} m \frac{\frac{1}{4} \sqrt{\frac{\pi}{b^3}}}{\frac{1}{2} \sqrt{\frac{\pi}{b}}} = \frac{1}{4} \frac{m}{b} \qquad (8.53)$$

where we have made use of the following standard definite integrals

$$\int_0^\infty x^2 \exp(-a x^2) \, dx = \frac{1}{4} \sqrt{\frac{\pi}{a^3}}$$

$$\int_0^\infty \exp(-a x^2) \, dx = \frac{1}{2} \sqrt{\frac{\pi}{a}} \qquad (8.54)$$

Substituting the value of b from Eq. (8.45) into Eq. (8.53) yields

$$I_x = \frac{1}{2} kT$$

The integrals I_y and I_z may similarly be evaluated and they also work out to be $\frac{1}{2} kT$. Thus we see that the average kinetic energy associated with each translational degree of freedom is $\frac{1}{2} kT$. We shall show that this is indeed the case whenever the energy associated with any degree of freedom is a quadratic function of the variable specifying that degree of freedom. This is the case which is usually met in any real situation. We know, for example, that the translational kinetic energy is a quadratic function of velocity; the rotational kinetic energy is a quadratic function of the angular velocity; the potential energy of a harmonic oscillator is a quadratic function of the displacement from the mean position, and so on.

The energy of a molecule is, in general, a function of all the coordinates of the cells in phase space within which the phase point of the molecule is located. For a system having f degrees of freedom, f independent coordinates of position $q_1, q_2, \ldots, q_f$ (also called generalized coordinates) and f independent velocity coordinates $\dot{q}_1, \dot{q}_2, \ldots, \dot{q}_f$ are required to specify the phase point of a molecule in the $2f$ dimensional phase space. Knowing these $2f$ coordinates of position and velocity (or momentum) the motion of a system for any future time can be determined.

Consider a general coordinate q_1 associated with one of the degrees of freedom. Suppose that $v_{q_1}(=\dot{q}_1)$ is the corresponding velocity component and $\frac{1}{2}\alpha_1 v_{q_1}^2$ the kinetic energy associated with this velocity component (here α_1 is a constant such as the mass or the moment of inertia of the molecule). We consider only the molecules whose velocities lie between v_{q_1} and $v_{q_1} + dv_{q_1}$; v_{q_2} and $v_{q_2} + dv_{q_2}$; $\ldots$; and v_{q_f} and $v_{q_f} + dv_{q_f}$. These molecules satisfy Maxwell velocity distribution. For the energy associated with these molecules we may write

$$E = E' + \tfrac{1}{2}\alpha_1 v_{q_1}^2$$

where E' is the average energy corresponding to the other degrees of freedom. E' is independent of v_{q_1} so that when integrating with respect to v_{q_1} we may consider $e^{-E'/kT}$ as a constant. We then have for the average value of $\frac{1}{2}\alpha_1 v_{q_1}^2$, i.e., the average kinetic energy per molecule associated with the velocity component v_{q_1},

$$\bar{E}_{q_1} = \frac{\displaystyle\int \ldots \int \tfrac{1}{2}\alpha_1 v_{q_1}^2 \exp\left(-E'/kT\right) \exp\left(-\alpha_1 v_{q_1}^2/2kT\right) dv_{q_1}\, dv_{q_2} \ldots dv_{q_f}}{\displaystyle\int \ldots \int \exp\left(-E'/kT\right) \exp\left(-\alpha_1 v_{q_1}^2/2kT\right) dv_{q_1}\, dv_{q_2} \ldots dv_{q_f}}$$

$$(8.55)$$

Since the variables $v_{q_1}, v_{q_2}, \ldots, v_{q_f}$ are all independent the integrals in Eq. (8.55) may as before be regarded as the product of two integrals. We have

$$\bar{E}_{q_1} = \frac{\int \frac{1}{2}\alpha_1 v_{q_1}^2 \exp\left(-\alpha_1 v_{q_1}^2/2kT\right)dv_{q_1} \int \ldots \int \exp\left(-E'/kT\right)dv_{q_2}\ldots dv_{q_f}}{\int \exp\left(-\alpha_1 v_{q_1}^2/2kT\right)dv_{q_1} \int \ldots \int \exp\left(-E'/kT\right)dv_{q_2}\ldots dv_{q_f}}$$

(8.56)

The limits of integration for all the variables are from 0 to ∞.
The second integrals in both the numerator and denominator are
identical, therefore they cancel out and Eq. (8.56) becomes

$$\bar{E}_{q_1} = \frac{1}{2}\alpha_1 \frac{\int_0^\infty v_{q_1}^2 \exp\left(-\alpha_1 v_{q_1}^2/2kT\right)dv_{q_1}}{\int_0^\infty \exp\left(-\alpha_1 v_{q_1}^2/2kT\right)dv_{q_1}}$$

which using the standard integrals in Eq. (8.54) simplifies to

$$\bar{E}_{q_1} = \frac{1}{2}\alpha_1 \frac{\frac{1}{4}\sqrt{\frac{\pi(2kT)^3}{\alpha_1^3}}}{\frac{1}{2}\sqrt{\frac{2\pi kT}{\alpha_1}}}$$

$$= \tfrac{1}{2}kT$$

Thus the average kinetic energy associated with the velocity component v_{q_1} is $\frac{1}{2}kT$. The same result can be proved for other
velocity components also. Thus according to the Maxwell-
Boltzmann statistics the mean kinetic energy per molecule associated
with each degree of freedom is $\frac{1}{2}kT$. In general, *each degree of
freedom which enters quadratically into the expression for the energy
of the entire system contributes, on the average, $\frac{1}{2}kT$ to the energy.*
This is a modified statement of the celebrated law of Equipartition
of Energy of classical statistical mechanics.

8.9. APPLICATION OF THE EQUIPARTITION THEOREM TO THE SPECIFIC HEATS OF GASES

The equipartition theorem may be verified in a simple manner
by applying it to the calculation of the specific heats of gases. The
mean total energy of a molecule with f degrees of freedom is,
according to the equipartition principle, $(f/2)kT$. Thus the total

energy of N molecules is $(f/2)NkT$ or $(f/2)nRT$, where n is the number of moles and R the universal gas constant. Thus the energy per mole is $(f/2)\,RT$. This must be equal to the total internal energy U of the molecules, i.e.,

$$U = (f/2)\,RT \qquad (8.57)$$

The molar specific heat at constant volume is

$$C_v = \left(\frac{\partial U}{\partial T}\right)_v$$

which on using Eq. (8.57) simplifies to

$$C_v = (f/2)R \qquad (8.58)$$

The molar specific heat at constant pressure applying Mayer's relation is

$$C_p = C_v + R = (1 + f/2)\,R \qquad (8.59)$$

We now calculate the molar specific heats for monatomic, diatomic and polyatomic molecules.

Monatonic Molecules

The degrees of freedom for a monatomic gas are the three components of the translational velocity. The specific heats per mole from Eqs. (8.58) and (8.59) are for this case,

$$C_v = \frac{3}{2} R$$

$$C_p = \frac{5}{2} R$$

Hence $\quad \gamma = \dfrac{C_p}{C_v} = 1.66$

These values for C_p, C_v and γ have been experimentally verified from the lowest to the highest attainable temperatures.

Diatomic Molecules

As the model of a diatomic molecule, we take a rigid dumbbell. In addition to translational energy, the molecule has rotational energy too. The rotation may be resolved into components about three axes. The rotation about the axis passing through the line joining the two atoms does not contribute towards kinetic energy as the moment of inertia about this axis is zero. Hence there are only two additional rotational degrees of freedom for a diatomic gas, and we have

$$C_v = \frac{5}{2} R, \; C_p = \frac{7}{2} R; \; \gamma = \frac{7}{5} = 1.4$$

This relation is also in general valid at room temperatures. However, for certain cases especially for hydrogen, it has been observed that the specific heat tends to attain the value for monatomic gases as the temperature is lowered. This does not agree with the principle of energy equipartition.

Polyatomic Molecules

Molecules consisting of more than two atoms have three degrees of rotational freedom. If we consider such molecules as rigid (internal vibrations assumed to be absent), we have

$$C_v = 3R, \; C_p = 4R; \; \gamma = \frac{4}{3} = 1.33$$

This holds for all polyatomic molecules having no internal vibrations.

QUESTIONS AND PROBLEMS

8.1. Starting from the principles of statistical mechanics deduce the Maxwell-Boltzmann distribution law $n_s = (g_s/f) \, e^{-\beta E_s}$ where the symbols have their usual meaning. What are the basic assumptions used in the derivation?

8.2. Deduce the Maxwell's law of distribution of molecular velocities. Describe some experiment which leads to its experimental verification.

8.3. Establish the law of equipartition of energy in classical statistical mechanics. Explain the term 'degree of freedom' in this connection. How has the equipartition law been applied to calculate the ratio of the specific heats for monatomic, diatomic and polyatomic gases?

8.4. Show that the most probable and the root mean square speeds for a molecule of an ideal gas are given by $v_m = \sqrt{\dfrac{2kT}{m}}$ and $v_{rms} = \sqrt{\dfrac{3kT}{m}}$ respectively. Find the ratio v_m/v_{rms}.

(0.817)

[Hint: To obtain the most probable speed v_m, find the value of v for which the speed distribution function is a maximum; set its first derivative with respect to v equal to zero:

$$\frac{d}{dv}[4\pi N a^3 v^2 \exp(-bv^2)] = 0$$

or

$$\frac{d}{dv}[v^2 \exp(-bv^2)] = 0$$

The root mean square speed is obtained as

$$v_{rms} = \sqrt{\bar{v}^2} = \left[\frac{\displaystyle\int_0^\infty v^2 \, dn_s}{\displaystyle\int_0^\infty dn_s}\right]^{1/2}$$

8.5. Calculate the average speed $\bar{v}$ for a molecule of an ideal gas.

Find the ratio $\bar{v}/v_{rms}$. $\left(\sqrt{\dfrac{8kT}{\pi m}}, \ 0.921\right)$

8.6. Find the average reciprocal speed $(1/v)_{av}$ in terms of temperature for a gas obeying Maxwell distribution of velocities.

$(2/\sqrt{\pi} \, v_m)$

8.7. Is the most probable molecular energy in an ideal gas equal to $\frac{1}{2}mv_m^2$ where v_m is the most probable molecular speed? If not what is the expression for the most probable energy?

$(\frac{1}{2}kT)$

[Hint: The most probable energy is that value of the energy for which the energy distribution function is a maximum:

$$\frac{d}{dE}\left[2\pi N\left(\frac{1}{\pi kT}\right)^{3/2} \sqrt{E} \exp(-E/kT)\right] = 0$$

or $\dfrac{d}{dE}\{\sqrt{E} \exp(-E/kT)\} = 0]$

8.8. Calculate the probability that the speed of an oxygen molecule lies between 100 and 101 metres/sec at 200°K.

(6.11×10^{-4})

[Hint: The probability that a molecule possesses speed between v and $v + dv$ is given by

$$P(v)dv = 4\pi \left(\frac{m}{2\pi kT} \right)^{3/2} v^2 \exp\left(-mv^2/2kT \right) dv$$

Here $m = 32\,\text{amu} = \dfrac{32}{6 \times 10^{23}}\,\text{gm} = \dfrac{32}{6 \times 10^{26}}\,\text{kg}$

$v = 100\,\text{m/sec},\ dv = 101 - 100 = 1\,\text{m/sec}$

$k = 1.38 \times 10^{-23}\,\text{J/°K},\ T = 200°\text{K}]$

9. Transport Phenomena in Gases and Brownian Motion

9.1. TRANSPORT PHENOMENA IN GASES

The molecules of a gas possess mass, momentum as well as energy which they carry with them while moving from one region of the gas to another. The molecules, therefore, serve as carriers or transporters of these physical quantities. The transport of any of these quantities takes place from the region where there is surplus of that particular quantity to the region where there is a deficit of that quantity. In the equilibrium state, however, the rate of transport of any quantity across a given plane in any direction is exactly balanced by an equal transport of the same quantity in the opposite direction. The transport of energy, momentum and mass gives rise to the phenomena of thermal conduction, viscosity and diffusion respectively. These phenomena are collectively called transport phenomena.

9.2. MEAN FREE PATH

A molecule moving about in a gas continuously suffers collisions with the other molecules. As a result of collisions, the direction and speed of the molecule changes. However, between two successive collisions the molecule travels in a straight line with uniform velocity. The actual path of a molecule in a gas after it has suffered a large number of collisions is therefore zig-zag, the path of the molecule changing after each collision.

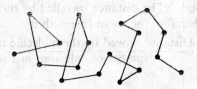

Fig. 9.1. Zig-zag path of a molecule in a gas.

The distance moved by a molecule between two successive collisions (which is a straight path) is called the free path of the molecule. It is convenient to define a mean free path for the general zig-zag path of the molecule. If S be the total distance travelled by the molecule in N collisions, then the mean free path is defined as

$$\lambda = \frac{S}{N}$$

A molecule can only interact with a neighbouring molecule provided the centre-to-centre distance between the two molecules is equal to σ, the diameter of the first molecule. Let us now assume that all the molecules with the exception of a single molecule are at rest. Let this molecule move with velocity v. This molecule in one second will pass through the centres of all those molecules which lie within a cylinder of radius σ and length v, having the volume $\pi\sigma^2 v$. If n be the number density of molecules then the number of molecules lying within this cylinder is $\pi\sigma^2 v n$. This must be equal to the number of collisions made by the mole-

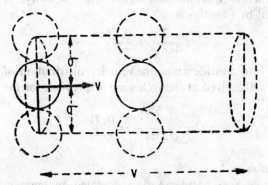

Fig. 9.2. **Diagram** showing a moving molecule. This molecule sweeps out a cylinder of volume $\pi\sigma^2 v$ in one second.

cule in one second. The distance travelled by this molecule in one second is v. Therefore the mean free path is

$$\lambda = \frac{\text{Distance moved by the molecule in 1 second}}{\text{Number of collisions in 1 second}}$$

$$= \frac{1}{\pi\sigma^2 n} \tag{9.1}$$

Eq. (9.1) may be expressed as

$$\lambda = \frac{m}{\pi\rho\sigma^2} \tag{9.2}$$

where ρ is the density of the gas. It is seen from Eq. (9.2) that

$$\lambda \propto \frac{1}{\rho} \text{ and } \lambda \propto \frac{1}{\sigma^2} \tag{9.3}$$

Thus the mean free path of the molecules is inversely proportional to the square of their diameters and to the density of the gas. Thus smaller the size of the molecule greater is the mean free path. Further, since the density is directly proportional to the pressure (for constant temperature) and inversely proportional to temperature, λ is inversely proportional to pressure and directly proportional to temperature.

Eq. (9.1) for the mean free path is derived on the assumption that all the molecules except one are at rest. In reality, however, the other molecules are also in motion. A modified expression for the mean free path has been derived by Clausius assuming that all the molecules move with the same average velocity which is equal to their mean square velocity. The expression derived by Clausius is

$$\lambda = \frac{3}{4\pi\sigma^2 n} = \frac{0.75}{\pi\sigma^2 n} \tag{9.4}$$

By taking into consideration the velocity distribution of the molecules Maxwell arrived at the following expression for the mean free path:

$$\lambda = \frac{1}{\sqrt{2}\,\pi\sigma^2 n} = \frac{0.71}{\pi\sigma^2 n} \tag{9.5}$$

Determination of λ

The mean free path can be calculated by knowing the molecular diameter because m and ρ can be easily estimated. But direct determination of σ does not lead to very accurate results. Therefore, direct measurements do no give very accurate values for λ. So, λ is only indirectly obtained by measurements on the viscosity.

9.3. VISCOSITY

The phenomena of viscosity arises when relative velocity exists between different layers of the gas involving transport of momentum from the fast moving to the slow moving layer. Consider a gas flowing left to right on a horizontal plane XOY. The layer of the gas in contact with this plane is at rest and the velocities of the layers increase with their distance from the horizontal plane. Thus a velocity gradient exists along z direction.

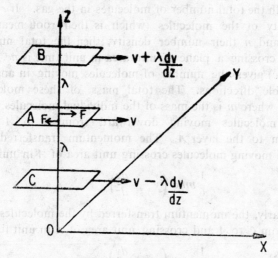

Fig. 9.3. Diagram showing three layers of a gas. A velocity gradient dv/dz exists in the z direction.

Consider A, B and C three layers of the gas (Fig. 9.3). The distance between the successive layers is assumed to be λ, where λ is the mean free path. If the intermediate layer moves with velocity v, the velocities of the layers B and C will be $v + \lambda \, (dv/dz)$ and $v - \lambda \, (dv/dz)$ respectively, because in the distance λ the velocity increases by $\lambda \, (dv/dz)$ for the layer B while it decreases by an equal amount for the layer C. The molecules are moving about indiscriminately in all possible directions in the gas in accordance with the kinetic theory. Since the molecules in the layer B possess greater momentum as compared to the molecules in the layer A, there will be a transference of momentum from B to A. This transference of momentum will be brought about by the molecules coming downward from the layer B and tending to cross layer A. The net effect of this momentum transport is to exert a tangential force F on the layer A resulting into the acceleration of this layer. The layer C will in the same manner exert a retarding force F on the layer A.

The molecules along any given axis can move in two possible directions. Considering all the three axes, the molecules can on the whole move in six possible directions. Therefore, the number of molecules moving in any direction along a particular axis will be one-sixth the total number of molecules in the gas. If c denotes the velocity of the molecules (which is their root mean square velocity) and n their number density, then the total number of molecules crossing a plane of unit area in unit time is nc. Therefore $1/6\,(nc)$ gives the number of molecules moving in any of the six possible directions. The total mass of these molecules is $1/6\,(mnc)$, where m is the mass of the individual molecules.

The molecules moving downward from B to A transfer momentum to the layer A. The momentum transferred by the downward moving molecules crossing unit area of A in unit time is

$$\frac{1}{6}\,mnc\left(v + \lambda\frac{dv}{dz}\right) \qquad (9.6)$$

Similarly, the momentum transferred by the molecules moving upward from C to A and crossing unit area of A in unit time is

$$\frac{1}{6}\,mnc\left(v - \lambda\frac{dv}{dz}\right) \qquad (9.7)$$

The net momentum transferred in the downward direction is therefore

$$\frac{1}{3}\,mnc\,\lambda\frac{dv}{dz} \qquad (9.8)$$

From Newton's second law this quantity must be equal to the viscous force F acting per unit area of A, i.e.,

$$\frac{1}{3}\,mnc\,\lambda\frac{dv}{dz} = F \qquad (9.9)$$

where
$$F = \eta \frac{dv}{dz} \tag{9.10}$$

η being the coefficient of viscosity.

Comparing Eqs. (9.9) and (9.10) we get

$$\eta = \tfrac{1}{3} mnc\, \lambda \tag{9.11}$$

We defer the interpretation of this equation for the time being and pass on to the discussion of the phenomenon of thermal conduction.

9.4. THERMAL CONDUCTION

Thermal conduction arises when the temperatures of the different regions of the gas differ from one another. The molecules transport thermal energy from the high temperature to the low temperature region tending to equalise the temperature difference.

We consider a gas confined between the two planes B and C perpendicular to the z direction at different temperatures. The plane B is at a higher temperature as compared to the plane C so

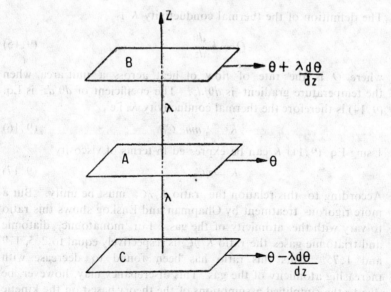

Fig. 9.4.   Gas confined between two planes B and C.   A temperature gradient $(d\theta/dz)$ exists along the z direction.

that in this case a temperature gradient $(d\theta/dz)$ exists instead of a velocity gradient.

We consider an intermediate plane A at temperature θ; the separation between the successive layers is λ. The temperatures of the planes B and C are $\theta + \lambda(d\theta/dz)$ and $\theta - \lambda(d\theta/dz)$ respectively. The molecules coming downward from the plane B to the plane A and crossing unit area of A in unit time carry thermal energy across this plane. The energy is

$$\frac{1}{6} mnc \ C_v \left(\theta + \lambda \frac{d\theta}{dz} \right) \tag{9.12}$$

where C_v is the specific heat of the gas at constant volume.

The thermal energy carried by the molecules moving upward from plane C and crossing unit area of A in unit time is similarly

$$\frac{1}{6} mnc \ C_v \left(\theta - \lambda \frac{d\theta}{dz} \right) \tag{9.13}$$

The net rate of flow of energy per unit area (which is identified with the rate of flow of heat per unit area) across the plane A is

$$\frac{1}{3} mnc \ C_v \lambda \frac{d\theta}{dz} \tag{9.14}$$

The definition of the thermal conductivity K is

$$Q = K \frac{d\theta}{dz} \tag{9.15}$$

where Q is the rate of flow of heat across a unit area when the temperature gradient is $d\theta/dz$. The coefficient of $d\theta/dz$ is Eq. (9.14) is therefore the thermal conductivity K, i.e.,

$$K = \tfrac{1}{3} mnc \ C_v \lambda \tag{9.16}$$

Using Eq. (9.11) K can be expressed in terms of viscosity

$$K = \eta C_v \tag{9.17}$$

According to this relation the ratio $K/\eta C_v$ must be unity. But a more rigorous treatment by Chapman and Enskog shows this ratio to vary with the atomicity of the gas. For monatomic, diatomic and triatomic gases the ratio $K/\eta C_v$ is respectively equal to 2.5, 1.9 and 1.75. Thus this ratio has been found to decrease with increasing atomicity of the gas. This discrepancy may, however, be due to the simplified assumptions of the theory based on the kinetic theory concepts.

9.5. VARIATION OF VISCOSITY AND THERMAL CONDUCTIVITY WITH TEMPERATURE AND PRESSURE

We shall now discuss the dependence of viscosity and thermal conductivity on temperature and pressure. Using Eq. (9.5) in Eq. (9.11) λ can be eliminated and we get

$$\eta = \frac{1}{3\sqrt{2}} \frac{m\,\bar{c}}{\pi\sigma^2} \qquad (9.18)$$

From this relation we see that the viscosity is independent of pressure. This fact has been verified experimentally for a wide range of pressure starting from a few millimetres up to several atmospheres. Also, as $\bar{c} \left(= \sqrt{\dfrac{8KT}{m}} \right)$ is directly proportional to $\sqrt{T}$ and inversely proportional to $\sqrt{m}$, η is directly proportional to $\sqrt{T}$ and $\sqrt{m}$ both. This viscosity of a gas therefore increases with temperature.

The above conclusions regarding dependence of viscosity on pressure and temperature apply to thermal conductivity as well. However, at low pressures both viscosity and thermal conductivity are not independent of pressure, but are rather proportional to pressure. Kinetic theory is able to account for this behaviour by extending the argument that since the mean free path becomes very large at low pressures, the assumptions of the continuity of the temperature and velocity gradients are no longer justified.

9.6. BROWNIAN MOTION

If very small, barely visible particles (colloid particles or pollen grains) be suspended in a liquid, the suspended particles exhibit an irregular vibratory motion as viewed under an ultramicroscope. This phenomenon first observed by an English botanist Robert Brown in 1827 is named after him as Brownian motion. The zig-zag, to-and-fro motion of the suspended particles is eternal, that is, the motion never seems to stop. Increasing the temperature of the solvent makes these haphazard motions more vigorous. The same thing happens when a less viscous liquid is chosen. Also, it is observed that only very small particles suspended in a liquid can exhibit this phenomenon. This is explained on the basis of the fact that in the case of an infinitesimal particle the impacts of the molecules of the liquid on it do not exactly balance at every instant, so that the particle is urged sometimes in one direction, sometimes in another.

The nature of this remarkable phenomenon, however, remained obscure for a long time. Only recently theories attempting to explain this phenomenon have been proposed. At first, explanation of this phenomenon was sought in the inhomogeneities of the temperature of the surrounding fluid or in other external influences. But the researches of Wiener and Gouy have made it clear that explanation must be sought in the collisions of the particles with the molecules of the surrounding fluid. In 1905, Einstein and Smoluchowski idependently developed a systematic theory of the phenomenon which has been experimentally checked by Perrin and others. The theory advanced by Einstein and Smoluchowski rests on the assumption that Brownian movement perpendicular to gravity is entirely irregular. But before giving an account of their theory we first give a simplified treatment of Brownian movement due to Langevin.

9.7. LANGEVIN'S THEORY OF BROWNIAN MOTION

According to Langevin the force on a suspended particle may be considered to be of two kinds, of which the first is a frictional force proportional to the velocity of the particle while the second incorporates the combined effect of all other external influences of the surrounding fluid. The frictional force, in particular, may be due to the frictional resistance (identifiable with the viscosity of the solvent) to which the suspended particle is subjected.

Consider the motion of a particle of mass m in any specified but arbitrary direction which we take as the x direction. Then, according to Langevin, we can write an equation of the form

$$m\ddot{x} = -f\dot{x} + X \qquad (9.19)$$

where $-f\dot{x}$ is the x-component of the frictional force and X is the combined effect of all other influences.

Multiplying through by x, Eq. (9.19) becomes

$$m\ddot{x}x = -f\dot{x}x + Xx \qquad (9.20)$$

which can be put in the modified form as

$$\frac{1}{2}m\frac{d}{dt}\left[\frac{d(x^2)}{dt}\right] - m\dot{x}^2 = -\frac{1}{2}f\frac{d}{dt}(x^2) + Xx \qquad (9.21)$$

If we form similar equations for all the suspended particles in the fluid and take the average of these expressions for all particles, we can write

$$\frac{\overline{m}\,d}{2\,dt}\left[\frac{d}{dt}(x^2)\right] - \overline{m\dot{x}^2} = -\frac{f}{2}\frac{\overline{d(x^2)}}{dt} + \overline{Xx} \qquad (9.22)$$

We now assume that since the force X varies in a completely random fashion, the average value of $\overline{Xx}$ vanishes. Looking upon the suspended particle as a large molecule its average kinetic energy must be equal to $\frac{1}{2}kT$ according to the law of equipartition, and so we have from Eq. (9.22)

$$\frac{m\,d}{2\,dt}\left[\frac{d}{dt}(\overline{x^2})\right] + \frac{1}{2}f\frac{d}{dt}(\overline{x^2}) = kT \qquad (9.23)$$

where we have made use of the fact that

$$\frac{d}{dt}(x^2) = \frac{d}{dt}(\overline{x^2}) \qquad (9.24)$$

Eq. (9.23) may be rewritten as

$$\frac{m}{2}\frac{du}{dt} + \frac{1}{2}fu = kT \qquad (9.25)$$

where

$$u = \frac{d}{dt}(\overline{x^2}) \qquad (9.26)$$

The general solution of the differential equation (9.25) is

$$u = kT\frac{2}{f} + C\exp(-ft/m) \qquad (9.27)$$

where C is a constant of integration. Since the mass m of the suspended particle is very small, the quotient f/m is very large and so the exponential term in Eq. (9.27) has no substantial influence after the first extremely small time interval. Hence

$$u = \frac{d}{dt}(\overline{x^2}) = kT\frac{2}{f} \qquad (9.28)$$

Integration with respect to t between the limits 0 and τ gives

$$\overline{x^2} - \overline{x_0^2} = kT\frac{2}{f}\tau \qquad (9.29)$$

If the initial displacement x_0 be taken to be zero, we have

$$\overline{x^2} = kT \frac{2}{f} \tau \tag{9.30}$$

Now since $\overline{x^2}$ which gives the mean square displacement is small, we can set $\overline{x^2} = \overline{(\Delta x)^2}$ and write Eq. (9.30) in the form

$$\overline{(\Delta x)^2} = kT \frac{2}{f} \tau \tag{9.31}$$

$\overline{(\Delta x)^2}$ in Eq. (9.31) is obtained by taking the mean of the squares of the displacements suffered by the particle in the x direction at equal time intervals τ. If $\Delta x_1, \Delta x_2, \Delta x_3 \ldots$ be the respective displacements suffered by the particle between successive intervals $0, \tau, 2\tau, 3\tau \ldots$ (i.e., Δx_1 is the displacement during time interval between 0 and τ, Δx_2 is the displacement between time intervals τ and 2τ, and so on), we can write

$$\overline{(\Delta x)^2} = \frac{(\Delta x_1)^2 + (\Delta x_2)^2 + (\Delta x_3)^2 + \cdots}{p}$$

where p is the number of observations on the particle starting from the initial time $t = 0$.

It should be noted that $\overline{(\Delta x)^2}$ is only loosely related to the true path of the particle so that it is not possible for us to analyse the actual motion of the particle which is indeed very complicated. Therefore, we have to contend ourselves with the description of only approximate rather than true motion of the particle.

Let us now imagine the frictional force f to arise from the viscosity of the solvent in which the particles are suspended. Applying Stoke's law the viscous force acting on a particle of radius a moving with velocity v in a fluid of viscosity η is

$$F = 6\pi\eta a v \tag{9.32}$$

The frictional force f which has the significance of a force per unit of velocity is therefore

$$f = F/v = 6\pi\eta a \tag{9.33}$$

Combining this equation with Eq. (9.31) yields finally

$$\overline{(\Delta x)^2} = kT \frac{\tau}{3\pi\eta a} \tag{9.34}$$

This relation is the same as derived by Einstein and Smoluchowski. We shall give their derivation in the next section. Eq. (9.34) has been experimentally verified by Perrin, Svedberg, Westgren, and others. Their results all agree in proving that $\sqrt{\overline{(\Delta x)^2}}$ is proportional to $\sqrt{T}$ and inversely proportional to $\sqrt{\eta}$. However, the temperature effect is not very marked for although $\sqrt{\overline{(\Delta x)^2}}$ is proportional to $\sqrt{T}$ it is also proportional to $1/\sqrt{\eta}$. Therefore, the pure temperature effect is outweighted due to the rapid decrease of the viscosity of the (liquid) solvent with increasing temperature. One also sees from Eq. (9.34) that $\sqrt{\overline{(\Delta x)^2}}$ is independent of the mass of the particle. This prediction of the theory has also been tested by Perrin. Perrin obtained from his measurements the value 6.85×10^{26} molecules/kilomole for the Avogadro's number. Westgren obtained the value 6.04×10^{26}. Both these values are very close to the now accepted value for the Avogadro's number.

Thus the experiments of Perrin and others led to the verification of Einstein's theory providing a convincing proof for the existence of molecules. Further, since molecular collisions are responsible for Brownian motion this also leads to the verification of the kinetic theory.

9.8. EINSTEIN'S THEORY

Langevin's theory does not throw much light on the physical nature of the phenomenon because it basically neglects the diffusion of the Brownian particles into the medium. We shall now give a simple treatment due to Einstein and Smoluchowski which takes account of the diffusion phenomenon. Diffusion takes place due to the unequal concentrations of the suspended particles at different parts of the medium. If we imagine a density gradient to exist, say in a horizontal direction which is taken along the x axis, the number of particles v crossing unit area of a vertical plane (Fig. 9.5) in unit time along the positive x direction is

$$v = -D\frac{dn}{dx} \tag{9.35}$$

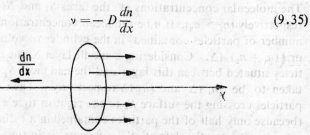

Fig. 9.5. Particles crossing a vertical plane opposite to the direction of density gradient.

where D is called the coefficient of diffusion. Negative sign is taken in Eq. (9.35) because diffusion of particles takes place in a direction opposite to one in which the density gradient exists.

Einstein calculated the diffusion coefficient in two ways. First, by considering the irregular motion of the suspended particles at right angles to gravity and second, by calculating the difference of the osmotic pressures between different parts, due to the unequal concentrations of the suspended particles. Einstein thus obtained two expressions for the diffusion coefficient which he equated to arrive at his result.

We first calculate the diffusion coefficient D from the random molecular motion. For simplicity we assume that each particle suffers the displacement Δ (this in fact is equal to $\sqrt{\overline{(\Delta x)^2}}$ of the preceding section) in the time τ. Imagine a cylinder of length Δ

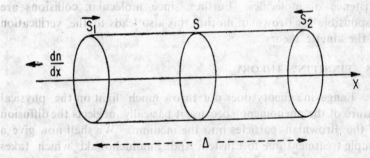

Fig. 9.6. Cylinder of volume $A\Delta$. Density gradient exists along the $-x$ direction.

and cross-sectional area A with its axis lying along the x direction. Its end faces are denoted by S_1 and S_2. Let a density gradient dn/dx exist along the axis of the cylinder, in the $-x$ direction. The molecular concentrations at the faces S_1 and S_2 are n_1 and n_2 respectively ($n_1 > n_2$). If n be the mean concentration then the total number of particles contained in the cylinder of volume $A\Delta$ is $nA\Delta$ or $\frac{1}{2}(n_1 + n_2)A\Delta$. Considering a middle layer S, the number of particles situated between this layer and the end faces S_1 and S_2 may be taken to be $\frac{1}{2}n_1A\Delta$ and $\frac{1}{2}n_2A\Delta$ respectively. Then the number of particles crossing the surface S_1 to the right in time τ will be $\frac{1}{2}n_1A\Delta$ (because only half of the particles contained in a cylinder of volume $A\Delta$ situated to the left of S_1 will cross it in time τ) and those crossing S_2 to the left will be similarly $\frac{1}{2}n_2A\Delta$. Therefore the excess of particles crossing S to the right is $\frac{1}{2}(n_1 - n_2)A\Delta$. From the definition of the diffusion coefficient (Eq. (9.35)) we have

$$\frac{1}{2}(n_1 - n_2)A\Delta = -D(dn/dx)\tau A \qquad (9.36)$$

But

$$-\frac{dn}{dx} = \frac{n_1 - n_2}{\Delta} \qquad (9.37)$$

So, we have from Eq. (9.36)

$$D = \Delta^2/2\tau \qquad (9.38)$$

We now calculate D in terms of the difference in the osmotic pressures. It has been shown by Van't Hoff that the Brownian particles act like the molecules of a gas. Therefore, all the gas laws are applicable to them. If the osmotic pressures exerted by the Brownian particles on the faces S_1 and S_2 be p_1 and p_2 respectively, then these are related to the molecular concentrations n_1 and n_2 by the relations

$$p_1 = n_1 kT, \; p_2 = n_2 kT \qquad (9.39)$$

where T is the temperature of the fluid and k is the familiar Boltzmann constant. The difference in the osmotic pressures, $p_1 - p_2$, gives rise to a force $(p_1 - p_2)A$ which is exerted on the particles contained in the cylinder. The action of this force is to push the cylinder to the right, i.e., along the positive direction of the x axis. The force acting on a single particle is

$$F = \frac{(p_1 - p_2)A}{nA\Delta} \qquad (9.40)$$

which on using Eq. (9.39) simplifies to

$$F = \frac{(n_1 - n_2)}{n}\frac{kT}{\Delta} \qquad (9.41)$$

Using Eq. (9.37), Eq. (9.41) may be expressed as

$$F = -kT\frac{1}{n}\frac{dn}{dx} \qquad (9.42)$$

This force must be equal to the viscous force acting on the particle as given by Eq. (9.33). We therefore have

$$F = 6\pi\eta a v = -kT\frac{1}{n}\frac{dn}{dx} \qquad (9.43)$$

where the particle has been assumed to be a sphere of radius a.
Eq. (9.43) gives

$$nv = -\frac{kT}{6\pi\eta a}\frac{dn}{dx} \qquad (9.44)$$

Comparing Eq. (9.44) with Eq. (9.35) we have

$$D = \frac{kT}{6\pi\eta a} \qquad (9.45)$$

since nv gives the number of particles moving to the right per second per unit area. Equating the two values of the diffusion coefficient as given by Eqs. (9.38) and (9.45),

$$\Delta^2 = \frac{kT\tau}{3\pi\eta a}$$

which is the same relation as derived earlier (cf. Eq. (9.34)).

QUESTIONS AND PROBLEMS

9.1. What do you understand by transport phenomena in gases? What are the various transport phenomena? Derive an expression for the mean free path of the molecules of a gas and show that it is inversely proportional to pressure and directly proportional to the temperature of the gas.

9.2. Obtain an expression for the viscosity of a gas in terms of its mean molecular free path, density and temperature on the basis of kinetic theory. Show that viscosity is independent of pressure but depends on the temperature of the gas.

9.3. Deduce from kinetic theory an expression for the thermal conductivity of a gas. Show that it is related to the viscosity η by the relation $K = \eta C_v$, where C_v is the specific heat at constant volume. How far does this relation agree with experiment?

9.4. What is the phenomena of Brownian motion? Give an outline of the Einstein's theory of Brownian movement. How has this theory been experimentally verified?

9.5. Calculate the mean free path, frequency of collision and the molecular diameter of nitrogen gas. Given $\eta = 166 \times 10^{-6}$ dynes-sec/cm², $c = 4.5 \times 10^4$ cm/sec and $\rho = 1.25 \times 10^{-3}$ gm/cm³. (Assume $n = 2.7 \times 10^{19}$ molecules/cm³.)
    (8.85×10^{-6} cm, 5.08×10^9 collisions/sec, 3.07×10^{-8} cm)

9.6. Calculate the pressure at which an oxygen molecule will have a mean free path equal to 15 cm. Assume the temperature of the gas to be 27°C, its molecular diameter as equal to 3×10^{-8} cm, Avogadro's number $N = 6.02 \times 10^{23}$ per mole and the gas constant $R = 8.32 \times 10^7$ erg mole^{-1} °C^{-1}.

(0.694 dynes/cm^2)

9.7. The viscosity of argon gas at N.T.P. is 210×10^{-6} dynes-sec/cm^2. Calculate (i) its molecular mean free path, and (ii) the number of collisions made per second. Given molecular weight of argon $= 40$ gm/mole, Boltzmann constant $k = 1.38 \times 10^{-16}$ erg/°C and Avogadro's number $N = 6 \times 10^{23}$ per mole.

((i) 8.57×10^{-6} cm, (ii) 4.8×10^9 collisions/sec)

Index